T0181933

Springer Geology

The book series Springer Geology comprises a broad portfolio of scientific books, aiming at researchers, students, and everyone interested in geology. The series includes peer-reviewed monographs, edited volumes, textbooks, and conference proceedings. It covers the entire research area of geology including, but not limited to, economic geology, mineral resources, historical geology, quantitative geology, structural geology, geomorphology, paleontology, and sedimentology.

More information about this series at http://www.springer.com/series/10172

G.S. Roonwal

Mineral Exploration: Practical Application

Springer

G.S. Roonwal
Inter-University Accelerator Centre
New Delhi
India

ISSN 2197-9545 ISSN 2197-9553 (electronic)
Springer Geology
ISBN 978-981-13-5440-3 ISBN 978-981-10-5604-8 (eBook)
DOI 10.1007/978-981-10-5604-8

Printed on acid-free paper

This Springer imprint is published by Springer Nature
The registered company is Springer Nature Singapore Pte Ltd.
The registered company address is: 152 Beach Road, #21-01/04 Gateway East, Singapore 189721, Singapore

Preface

Geologists and mining people are concerned mostly with the geological aspects of mineral resources and required to be able to judge the technical feasibility and the estimate of economic potential of a mineral deposit when recommending a defended exploration. We need to recognize the significance of geological, technical, administrative and political influence on mining and mineral processing when evaluating economic potential of a prospective deposit. Mining engineers here have to understand the ways in which geological, economic and political factors affect the feasibility of a project.

In this book, I wish to introduce students, beginners and possibly professionals to the essentials of concepts of mineral exploration, mine evaluation and resource assessment of the discovered mineral deposit. It is true that each of these aspects is available in detail in the specialized literature. I have attempted to integrate their important aspects with an aim to give them to those who are willing to get involved in mineral exploration and resources with broad synthesis.

As we have the knowledge of the earth crust, structure and in particular the regional setting of ore deposits have no doubt increased during the past decades. But much of it is concerned with petrology, mineralogy, structural and stratigraphy. Little is on the exploration of mineral deposits. When present, it is in papers scattered through the journals. The Society of Economic Geology, a geochemical exploration society, has added to our knowledge of mineral exploration. It is thus important to present a book hard for adoption for starting serious interest in exploration of ore deposits.

The task of compiling this book proved to be a formidable one. To achieve this, I needed the support and cooperation of several people. I should acknowledge the direct support so readily and cheerfully provided. A ready and cheerful support came from my colleagues at the Inter University Accelerator Centre (IUAC), and I should specially mention my sincere thanks to Dr. D. Kanjilal, the Director of IUAC, and Dr. Sundeep Chopra, Head of our AMS Unit. They both cheerfully supported my endeavour. However, this has become possible only due to support of the Department of Science and Technology (DST), and the in-charge of USERS scheme Dr. S.S. Kohli in particular.

This compilation would not have been possible without the willing discussions and encouragement of professional colleagues. Professor Hojatollah Ranjbar of Shahid Bahonar University of Kerman kindly read through the full manuscript and contributed a chapter on remote sensing in mineral exploration and Mr. S.N. Sharma of Faridabad for review of surveying chapter and Prof. B.C. Sarkar of the Indian School of Mines (ISM), Dhanbad and Prof. A.K. Bansal of Delhi University, for the review of chapter on geo-statistics in mineral exploration. And I am pleased to acknowledge the rapid and efficient assistance of Narender Kumar; Mr. Aninda Bose and Ms. Kamiya Khatter of Springer, New Delhi, for their support. At home, I have received encouragement from Prashant, Geetu and loving care of Kabir and Kanav during the writing of this work.

After all mineral exploration is like: *"A child lies in a grey pebble on the shore until a certain teacher picks him up and dips him in the water, and suddenly you see all the colours and patterns in the dull stone, and it's marvellous for the teacher"* (Elizabeth Hay in Alone in the Classroom, McClelland and Steward 2011, p.94).

New Delhi, India G.S. Roonwal

Contents

About the Author

Prof. G.S. Roonwal has been formerly Professor and Head of the Department of Geology, University of Delhi and Director, Centre of Geo-resources, University of Delhi South Campus. He is at present Honorary Visiting Professor in the Inter University Accelerator Centre, an autonomous Institute of UGC-MHRD. He is a prominent geoscientist of international stature. He is a winner of the National Mineral Award of the Ministry of Mines, Government of India, and has received several awards and medals by the professional societies. He has worked in Europe and North America in several well-known institutions and participated in oceanographic expeditions.

Professor Roonwal has made significant contribution in land and marine geology which includes exploration assessment and eventually mining and environmental sustainability of land and sea-floor mineral deposits such as volcanogenic massive sulphides which contain valuable metals such as zinc, copper and associated gold and platinum. He also authored books on land and sea-floor mineral resources. His books such as Mining and Environmental Sustainability (2014), Iron Ore Deposits and Banded Iron Formation of India (2012), The Story of Science from Antiquity to the Present (Geology Section) (2010), The Indian Ocean Exploitable Mineral and Petroleum Resources and edited books entitled "India's Exclusive Economic zone" (with Dr. S.Z. Qasim), Living Resources of India's Exclusive Economic Zone" (with Dr. S.Z. Qasim) and Mineral

Resources and Development (2005) (Roonwal et. al.). He has written a number of scientific papers published in established journals. Professor Roonwal is listed in several biographical citations such as Marquis Who's Who in the World, since 1999 and Marquis Who's Who in Asia since 2010.

Chapter 1
Mineral Resources and Exploration

1.1 Concept of Mineral Exploration

The aim of the mineral exploration is to discover deposits of such size, shape, and grade and in such place that they can be profitably extracted from the earth's crust. The deposits need to be reckoned in a way that the mineral and products on which modern society is dependent are usefully utilized. An understanding and insight into clues that guide these efforts in locating a deposit are many and varied. They would range from details about the composition and texture of specific rock type, or form of a particular rock body to large-scale hypothesis on the composition of earth's more than 4 billion years ago, and possible influence on the chemical procedure effecting mineral deposition. Society need minerals for (i) Life support, soil, salt and fertilizers, (ii) Energy generation, coal, oil, gas and nuclear minerals, (iii) Materials for construction of shelters, industrial plants and machines; and for manufacturing appliances and commercial/industrial products.

The tools used in exploration are equally diverse: topography, geological, geophysical and geochemical maps at many scales and description or interpretation models of individual deposits or areas, or regions of ore body types and the metallogenic provinces and epochs. The aim of the handbook is not to discuss world economic geopolitics of minerals, the demand for access, domestic environmental problem, and legislation, supply and demand of both strategic and non-strategic minerals, or the human and financial resources of a mining company, its corporate goals or areas of expertise. We know that these factors are critical to the scope and timing of exploration program, and to the selection of prospecting regions of target, size and type. These are considerations that limit and restrict both the activities and the thoughts of the exploration geologist. Therefore in a true sense, a geologist has to work within the variable parameters in his endeavours.

Let us understand that ore bodies are the product of one or more cycles of concentration, by a variety of geological processes of a particular element or a group of elements. As a result ore deposits are extremely rare geological features.

G.S. Roonwal, *Mineral Exploration: Practical Application*,
Springer Geology, DOI 10.1007/978-981-10-5604-8_1

They are distributed sparsely within the earth. To the ancients, their occurrence was random, later came to the realization that certain types of ores were characteristic of particular geographic areas, and the concept has been steadily improving, but still has an incomplete geological understanding of gross distribution of ores and types of deposits in space and in time. Therefore, this handbook with practical guideline to the beginner needs to be accepted in the spirit of practice.

Examining the historical aspects, it emerges that prior to World War II, the literature of ore bodies was largely descriptive. They attempted to explain origin on general and speculative. Although much remain unknown, our knowledge of the formative pressure and temperature condition, the sources, volumes and flow rates of solution, the concentration of elements and isotypes within them is greater today. This knowledge led to much more rigorous and definitive concepts of ore genesis.

We need to, therefore, understand that it is always difficult and equally expensive to find ore bodies. The better our understanding of known deposits, the more efficient and rational our search for yet undiscovered ores will be. This quest for knowledge has to contain both in the field and in the laboratory. It has to be regional as well as local, genetic as well as descriptive. The general problem needs to be solved in the following way:

1. What are sources of the metals, water, the halogen and alkali and sulphur?
2. What are the “mechanisms” and conditions of mobilization, transport and concentration of different element and compounds?
3. What controls deposition?

For purpose of the handbook, shall be concerned only with those more than 60 minerals that occur in solid form as discrete bodies, mineral deposits, but not as common natural rock formation.

Mineral deposit is physically exhaustible and thus non-renewable. The mineral extracted from it are non-renewable, but some of them, especially the “materials” minerals are also increasingly reusable. Also, other commodities may be substituted for most of the “materials” minerals to satisfy the same human needs. The old perception of minerals as completely depletable is gradually being modified as society realizes that throughout the history non-renewable mineral reserves and recyclable stocks have constantly been expended more readily than renewable resources.

In describing mineral resources, a grouping shown as (Fig. 1.1) despite reservation on the use of term “resources” is a practical approach. There is need for improvements through research since the term resources need to be converted to reserves and reserve to minable quantity. This requires: (i) An understanding of the formation of mineral deposit, (ii) Discovery-oriented exploration for new deposits, e.g. creation of additional resources through discovery of new deposits.

Because of our inquisitive efforts, sustained curiosity and ingenuity, it is reasonable to assume that in the coming years we shall find new ways to meet needs—such which are unrecognized today will become evident with time. This optimistic view of the future is a challenge and responsibility. It gives opportunity for us to

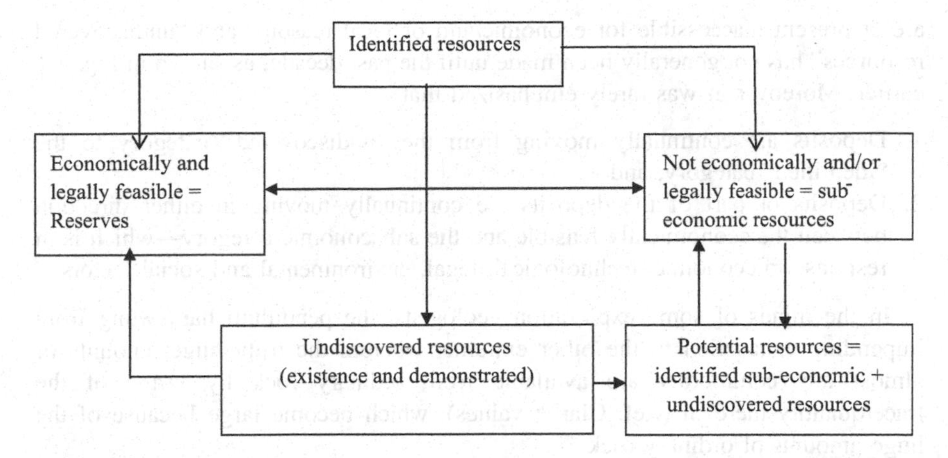

Fig. 1.1 Relationship between mineral resources and reserves (Modified after Bailly 1978)

prepare for the future on the basis of our requirements and the resources we perceive and aims for new sources, new opportunities and new alternatives.

1.2 Accessible and Available Resource and Reserves

In a free market economy, a mineral deposit from which usable minerals or commodities can be economically and legally extracted at time of determination is known as "reserve". It fits the popular concept of a resource that thing on which one can turn to when in need, and it also fits in economists concept of resource that it can actually or potentially create new wealth and its value is reflected in market. However, "reserves" do not cover all mineral concentration of interest and we, geologists, need to find a broader definition of "resources" to be used, as explained in an earlier figure.

The controlling factors sometime known as Five 'E's that determine whether a mineral deposit is a reserve are:

1. Existence of the deposit.
2. Extractability of the mineral value
3. Energy and material required for extraction
4. Environmental acceptability
5. Economics of possible operation including capital availability and cost, operating cost, market demand, and societal, fiscal and legal constraints.

These aspects and guidelines proposed are given in different locations within the following chapters of this handbook.

The distinction between "Ore reserves" known material economically and legally feasible activity, "identified resources" ore reserves plus known materials that are at present inaccessible for economic/and or legal reasons, and "undiscovered resources" has not generally been made until the past decade, as shown in Fig. 1.1 earlier. Moreover, it was rarely emphasized that

1. Deposits are continually moving from the "undiscovered" category to the "identified" category, and
2. Deposits or part of the deposits are continually moving in either direction between the economically feasible and the subeconomic category—which is in response to economic, technological, legal, environmental and social factors.

In the minds of some exploration geologists, the pendulum has swung from impending exhaustion to the other extreme, wherein the truly huge amounts of almost any commodity are available from ordinary rock by virtue of the trace/quantity there in (Ref. Clarke values), which become large because of the huge amounts of ordinary rock.

The latter opinion has captured the minds of some economists, who insist that there will never be a mineral exhaustion because increasing prices will increase the size of "reserve" box thereby increasing supply. However, this is disputed for primarily two main reasons:

1. Costs are measured not simply in term of money, but also in terms of energy, manpower, other natural resources, environmental degradation/restoration and such other aspects.
2. Once the mineralogical threshold is reached, there is an immense and universal increase in the cost and difficulty of extracting the metal—so great that the rock is no longer a practical source (such as seafloor "nodules").

The concept that resources are properly measurement in economic, not physical term is an economists' point of view that appears to be valid over the short term and most applicable to "resources" estimates. In this way, we can be confident of an adequate supply of practically all mineral commodities in the coming decade. However, the price mechanism concept does not reflect the physical limitation of the earth's crust, the requirement for gangue or waste disposal, the unpredictability of time of discovery, the availability of capital, equipment, and above all manpower, the time, political and social constraint on development and production, including access to the deposit and the possible related scarcity in the future. There is an ultimate limit to resource that is reached when the total investment of materials and energy required to procure a commodity exceed its values to society.

1.3 Estimation of Undiscovered Resources

Paucity of reserves has always been used to indicate the need for additional research on extraction technology and/or additional exploration for the new deposits. The necessity to improve planning for the future, however, and to avoid scarcities, has caused many to try to estimate both the likelihood of extraction technology breakthrough and the amount of commodities that may exist in general, for discovery and eventual transformation into reserves or subeconomic resources.

It has been expressed in estimates by geo-statistician and geologists of the quantities of mineral that may be found in yet undiscovered deposits: the "undiscovered resources" is now applied to things whose very existence is explained by extrapolation, in contradiction of the popular concept of a resources—a thing that can be counted on when in need.

Perhaps undiscovered resources need to be called as "postulated deposits" while "reserves" should be retained with its accepting meaning, and the phrase "resources should not be used to included postulated deposits". Nevertheless, for policy formulation and for exploration planning, government and the mining industry, respectively, are interested in developing estimates of undiscovered resources. Geo-statistical and geologic approach have been used. Many geo-statistical techniques have been applied, but most do not arrive at results usable in exploration because they develop mainly global or regional numbers for total undiscovered resources without indicating which piece of lands is more likely to contain a mineral deposit (see Chaps. 3 and 6).

Some of these geo-statistical approaches have been described and evaluated in a detailed way in Chap. 6. The geologic approach in its varied forms follows steps that are similar to those used in exploration planning. Thus, the geo-statistical approach has some value for national or global planning, and formation of mineral policies, whereas the discovery-oriented approach has immediate application in exploration. The former tries to foresee the future, the latter tries to enable it. However, no approach can even yield results superior to the postulates used to formulate the predictive model.

1.4 Economic—Definition of Exploration

Exploration, the search for yet undiscovered deposits, is justified for some commodities

(a) The need for new reserves when known reserves are inadequate for the foreseeable future.
(b) The desire to find deposits that will, it is hoped, be more profitable (lower cost and higher grade) then those currently mined or held as reserves.

This is illustrated in Fig. 1.2.

The mineral industry, to assure its viability, needs to make decisions about new deposits and old deposits under time tested criteria of return on investment. Hence, it is important for industry to pursue work only on those projects that continue to meet acceptability criteria in the project proposals through the four main steps in a successful exploration program.

Exploration Effectiveness
$EFT = \frac{\text{Value of discoveries}}{\text{Expenditure for all exploration projects}}$
Project Effectiveness
$PE = \frac{\text{Value of discovery}}{\text{Expenditure for projects}}$
Success Ratio
$SR = \frac{\text{Expenditures for successful project}}{\text{Expenditure for all projects}}$
$EFT = PE \times SR$

Fig. 1.2 Nature of successful metallic minerals exploration project (based on Bailly 1978)

1. Regional reconnaissance
2. Detailed reconnaissance for favourable area, then to
3. Detailed surface appraisal of target area, and finally to
4. Detailed three-dimensional sampling and feasibility study.

For industry, the project turns negative—when the expected rate of return falls below the minimum acceptable rate of return. This minimum rate is used as the discount factor in calculating the present worth.

The exploration geologists of old tactical style developed and handed down through generation of empirical observations, "Ore was where we found it". During the early era of exploration and prospecting, geologists were not involved in mineral exploration. In determining where to seek next discovery, the exploration geologists used predictive tools which have been improved at an accelerated rate in the past few decades. Today exploration geologists take for granted some of the most useful and commonly used ore occurrence and ore genesis models. However, in doing so, one needs to keep in mind that their heritage may be ancient, but their useful forms are recent.

Our ability to prepare useful mineral occurrence prediction maps—the basis for exploration planning—is continuously improving. In preparing such maps, we use models that describe the conditions that appear to be necessary for the formation of deposits. However, we also know that even these valuable tools are not sufficient to guarantee ore deposits discovery.

A. First generation discoveries were made through empirical observation by prospectors, or/and exploration geologists and
B. Second generation discoveries are made through technical surveys in the vicinity of previous discoveries.
C. For several types of deposits, third generation of discoveries that are essentially conceptual based on scientific identification of area where ore is predicted to occur.

The outlook is for increased confidence in the validity of predictions about mineral occurrences. This new confidences can be further improved only through

Table 1.1 Application of geology and relationships between geology and mineral exploration (after Bailey 1978: Scientific concepts models and its applications)

Stages of exploration and development	Observation and documentation		Association in nature		Models/hypothesis/ theories	
	Deposits	Regional maps	Ore pattern	Non-genetic model	Genetic model	Geological concepts
Regional appraisal	Minor	Major	Major	Major	Major	Minor
Detailed regional targeting	Major	Major	Major	Major	Moderate	–
Detailed surface appraising of target	Major	Minor	Major	Major	Moderate	–
Detailed 3D evaluation of target	Moderate	Moderate	Moderate	Minor	–	–
Mining—exploration	–	–	Minor	Minor	–	–
Metallurgical—recovery	–	–	Minor	Minor	–	–

continued research on ore deposits. Our knowledge of the geology of mineral deposits—usable in exploration—consists of the following order of increasing generalization (Table 1.1).

1. Factual description is based on the observation of deposit and their settings, available through published geological maps, recorded memoirs, etc. These existing map and published information are used as the main source of target concepts in today's exploration. The significance of these maps and published information is of great help for their contribution in planning of our effective exploration.
2. Recognized mineral associations and metal association will help to point pattern to look for in exploration. They will guide to define signs of possible preserve of valuable deposits.
3. Minerals deposit occurrence models that integrate all the metal association with certain types of deposits. The target hypotheses derived from such models are generators in the minds of the explorer, who make a case for the postulate that ore may occur or be found where he predicts it to occur.
4. Genetic models give an academic satisfaction and an explanation of depositional environments and deposition mechanism that are involved in creating certain types of deposits.

5. Geological laws of nature or theories/concepts of general scope. Though these are of general in nature, they may still be helpful in regional appraisal of a deposit bearing area.

On a general way, any mineral exploration may affect exploration result only 3 or 5 years after its concept formulation. It may take several years to move from preliminary to a reliable assessment. It is general experience that it takes 5 years, on an average, to conclude a successful exploration project with the discovery of a reserve, and then it takes up to 7 years to get the deposit ready for mining.

Mineral exploration is time consuming, geological research shall become important factor in converting resources into reserves, and future production, but we need to understand it is a very slow acting process. Because of its great potential value to mankind, geological investigation therefore needs more support, it needs patience and sustained environmental support in locating new deposits.

1.5 Geological Cycles and the Formation of Mineral Deposits

We understand that the earth has evolved to its present configuration of a cool, solid crust carrying interior, partially molten from a planet with molten, homogeneous crust about 4 billion years ago. The processes involved in the cooling, crystallization and reworking state of the primondial earth's materials are quite similar to what we see taking place in at present time. We understand that there must have been such geological events, these processes are so very slow so that form mineral deposits. However, geological processes taking place at present at a rate so slow that only accumulation formed over a long geological time period that it was concentrated enough to be exploitable in the present or near future. In this way, we understand that mineral deposits are non-renewable in relation to human time frame.

Therefore, we can say that ore-forming processes and ore deposits are closely related to rock association, geological settings and geologic cycles that we attempt to understand.

We know that deep in the earth's crust, magma upwells and crystallites as it cools do it pushes to reach to surface. A surface expression of such magma activity is seen in the intense volcanic activities that we observe at present time as well such as in Hawaiian islands and elsewhere. We also know and observe the existing mountain ranges when erode, and water carries the clastic and chemical constituents to depositional basins in which sedimentary deposits of all types—gravel, sand, clays, limestone and other get deposited. Down wrapping of the earth's crust under these basins lead to burial ofsediments to depth at which they are re-crystallized and metamorphosed to give new rock types. Earth's structured movements cause local fractures in the crust—they provide channel ways for ore-forming solutions. This leads to precipitation of minerals deposits and ore deposits in veins form.

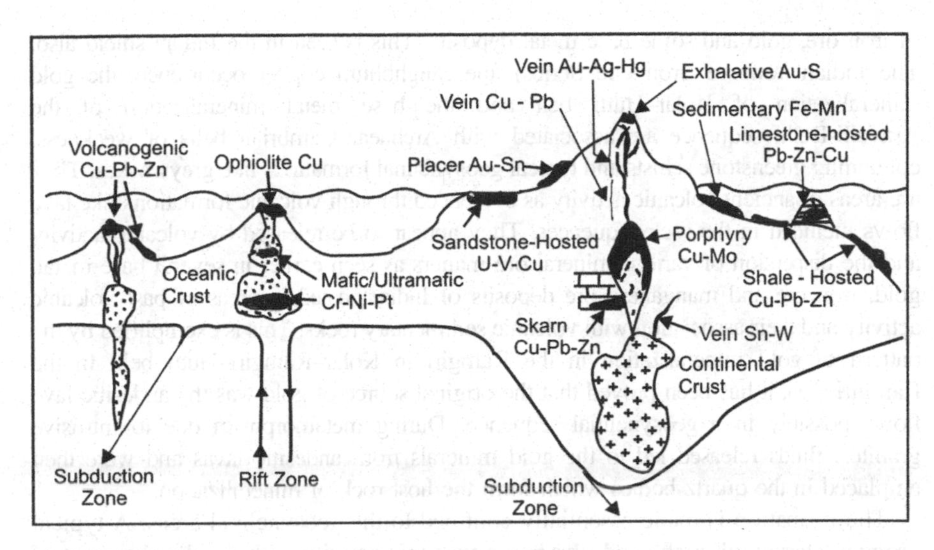

Fig. 1.3 Schematic diagram showing the geologic settings of mineral deposits (after Gocht et al. 1988)

These geologic rock forming and structural changes and developments have been explained through understanding of plate tectonic model of crustal evolution, as shown in Fig. 1.3.

The present tendency is to accept syngenesis as well as epigenesis as valid in their contexts but lays stress on the observational, geometric and geochemical nature of the deposit, for placing it in one group or the other. Broadly, all congruent deposits are syngenetic and all non-congruent deposits are epigenetic.

There are a several known examples that have shown the syngenetic as well as the epigenetic nature. The occurrence of diamonds within ultramafic is an example of the syngenetic concept. Clear-cut examples are the epigenetic deposits formed within fault planes and other structural features. But, between the two clear-cut instances, there are several examples that do not show to either one without contradicting field evidences. A large number of base metal, gold and other deposits particularly in the Precambrians show such problems which preclude their being as distinctly syngenetic or epigenetic.

Ore genesis concepts do not view the process of mineralization in isolated in the earth's history, but as a part of the major processes responsible for the emplacement of rocks and their tectonic evolution, suggests that there exists a close relationship between mineral provinces and the major belts of weakens and deformation in the earth's crust. The boundary between the Archaeans and the Cambrians forms such a belt of weakness in all shield areas of the world. They are known for the occurrence of iron ore, gold and some base metal deposits. This is seen in the Indian shield also. The Indian iron ore (Iron Ore Series), the Singhbhum copper occurrence, the gold mineralization of Kolar–Hutti belt and the base metal mineralization of the Aravalli-Delhi sequence are associated with Archaean-Cambrian belts of weakness, containing greenstone schists and typical geosynclinal formations like greywackes. They are areas of ancient volcanic activity as evidenced through volcanic formations like lava flows included in the rock sequences. They appear to be released by volcanic activity and the dispersion of various mineral ore matters as seen earlier in several base metal, gold, iron ore and manganese ore deposits of India located in areas of past volcanic activity and their association with volcanic sedimentary rocks. This is exemplified by the pattern of gold mineralization in the Ramgiri in Kolar–Ramgiri–Hutti belt. In the Ramgiri area, it has been proved that the original source of gold was the andesitic lave flows possibly in a geosynclinal sequence. During metamorphism due to intrusive granites, fluids released led to the gold minerals from andesitic lavas and were then emplaced in the quartz bodies which form the host rock of mineralization.

The formation is mafic essentially confined to the geosynclinal areas. A typical orogeny starts with mafic and ultramafic volcanic activity with the development of large areas of serpentinization at the bottom of geosynclinal troughs. The trough is progressively filled up with sediments and the geosynclines sink into the crust. The lower part of the geosyncline undergoes metamorphism, migmatization and granitization. Granites owing to their lower specific gravity rise within the geosyncline in the form of large plutons. Andesitic magmas rise to the surface in subsequent cycle of volcanism to stabilize the last phase of orogeny.

In the case of gold mineralization, mineral deposits tend to form initially within the volcanic rocks or near them in sediments originating from volcanic derivatives. The rise in temperature during the later part of orogeny shown by large-scale intrusion leads to metamorphism of geosynclinal rocks. The metamorphism produces ore fluids within the rock which cause ore migration, reconsideration, metamorphism of ore deposits. It causes wall rock alterations, metamorphism and replacement alkin to typical hydrothermal processes. Mineral deposits resulting from volcanic-geosynclinal association are called volcanic sedimentary-exhalative deposits or as exhalative sedimentary deposits.

Thus, doubt exists if all the exhalative sedimentary deposits can be considered syngenetic. Though two characteristics indicate this being as syngenetic; (i) They show clear lithological affiliations, a deposit will be confined to one suite or facies of rocks. (ii) They are invariably congruent with host rock suggesting syngenetic. Adopting this concept, a following set of criteria will be useful in recognizing exhalative sedimentary deposits:

(i) The syngenetic deposits are developed within definite stratigraphic zones,
(ii) Inter-banding of ores of contrasting composition and host rock will be common particularly when clear sedimentary host rocks are present,
(iii) Cross-cutting veins will be absent or very rare,
(iv) Ore minerals vary with variation in sedimentary facies when the host rock is a sedimentary unit,

(v) Metamorphism of the ore and the host rocks are isofacial and
(vi) Wall rock alteration may be absent.

Mineral deposits formed by the exhalative sedimentary process in typical geosynclinal environments can be grouped into five classes representing the stages of development of the geosynclinals series:

(i) **Early orogenic deposits**: These deposits that occur in association with spilitic lavas show serpentinization. Due to subsequent burial in deep piles of sediments, they generally occur in a metamorphosed condition. The mineralized horizon may occur under a suite of pillow lava rocks, indicating post-mineral subaqueous volcanic activity. The ore bodies are massive pyrite type with minor amounts of As, Zn, Pb and Ni. The minerals here are fine-grained pyrite, accompanied by chalcopyrite, spharlerite and hematite. Typical host rocks are spilitic pillow lavas, chert beds and mudstones.
(ii) **Syn-orogenic deposits**: Deposits of this type occur in association with typically high metamorphic rocks. The degree of metamorphism generally reflects the intensity of orogeny. Typical deposits show disseminated to massive sulphides with pyrrhotite as the principal mineral constituent. Chalcopyrite, cubanite and magnetite occur as subordinate minerals. Interbedding of the host rock with sulphides is common.
(iii) **Late orogenic deposits**: In the late orogenic stage, volcanism is acidic with rhyolitic lavas and acidic tuffs. The deposits of this type ore are of large dimensions unlike the earlier two types where ore generally tend to be rich in grade but small in dimension. Here, massive sulphide type with pyrite is the predominant ore. The associated sediments generally show low grade of metamorphism.
(iv) **Final orogenic deposits**: The three types of deposits described so far have all some common genetic features, deep geosynclinal association, greywacke sediments and pyrite or pyrrhotite forming the major ore mineral. The final orogenic phase is free of any volcanic activity. Here, sulphide mineral is pyritic but different mineral assemblages both with fairly abundant Fe, Zn, Pb, Ba sequence are also present in a distinctly calcareous or dolomitic. Biogenic agencies are thought to play a role in the mineralization. Black carbonaceous rocks are present in the sequence.
The Sargipali sulphide deposit bears close resemblance to the conditions described above. The host rock is a calcsilicate-quartzite. Mineralization is within a sequence of the Iron Ore Series with grantitic intrusive rocks. The mineral sequence shows galena, chalcopyrite, pyrite, pyrrhotite, arsenopyrite, tennantite, tetrahedrite and silver.
(v) **Post-orogenic deposits**: These deposits occur subsequent to the major orogenic phase with a marginal connection between this deposits and volcanism. They show more evidences of fumerolic and biochemical actions. A typical environment being enclosed coastal basins of lagoon type. The minerals are arsenopyrite, sphalerite, galena and barites (As, Fe, Cu, Pb, Zn, Ba).

The lead-zinc deposits of Zawar and Rajpura—Dariba may have such origin described above. However, they have been deformed so much that their original environment is no longer clearly discernible. Such deposits may have been remobilized in their original setting during the evolution of the geosyncline.

1.6 Project Effectiveness and Success Ratio

We have seen that very few of the chemical elements of which the earths' crest is composed of are present in elemental form; gold and sulphur are amongst these exceptions. Most are in minerals containing two or more elements. The elements are distributed irregularly and a few are much more abundant than the others, so that eight, e.g. oxygen, silicon, aluminium, iron, calcium, sodium, potassium and magnesium, account for over 98% of the crust.

Our particular interest in the different kinds of rocks and minerals depend upon their usefulness for many purposes. A mineral in the economic sense often can be a useful substance obtained from the earth's crust. Therefore, concentrations of elements as minerals, appreciably above the average of their abundance in earth's crust are known as mineral occurrence if they have no foreseeable value, but if the accumulation can be exploited now or at some time in the foreseeable future they are described as mineral resources, as shown in Fig. 1.4.

Minerals and mineral-based industries play a vital role in the economy of all nations. Since ancient times, many metals like iron, copper, lead and zinc have been extracted. As these metals occur mostly in the ore form, and since the very early days of civilization, man has been aware of the existence of ore concentrations mineral deposits in certain places. Although the extraction of metals and their uses are widely mentioned in ancient literature, there is very little reference to any organized attempts being made to locate them. However, it is reasonable to assume that the ancient people had devised some means of locating metallic and non-metallic ore deposits. The evidence of this can be seen in India at several places such as Udaipur area Zawar and RajpuraDariba mines in Rajsamand; and RampuraAgucha in Bhilwara area, in Khetri region at Bhogani and other parts of India. Although India has been in the field of mineral industry for long, it is only since the commencement of 5 years development plan that led to industrialization and consequently the demand for metals and ores started growing rapidly. Therefore, the need for locating new deposits has also started increasing, giving a fillip to the science of exploration geology.

In the earlier days, mineral exploration was done by individuals, known in the western countries as prospectors. Mineral discoveries depended upon the skill, diligence and persistence of these individuals. Though prospectors are active in countries like the USA and Canada, the times of the individual prospector are limited. Discoveries of new deposits especially those which are concealed or underground are getting scarce as unexplored areas are becoming smaller all the

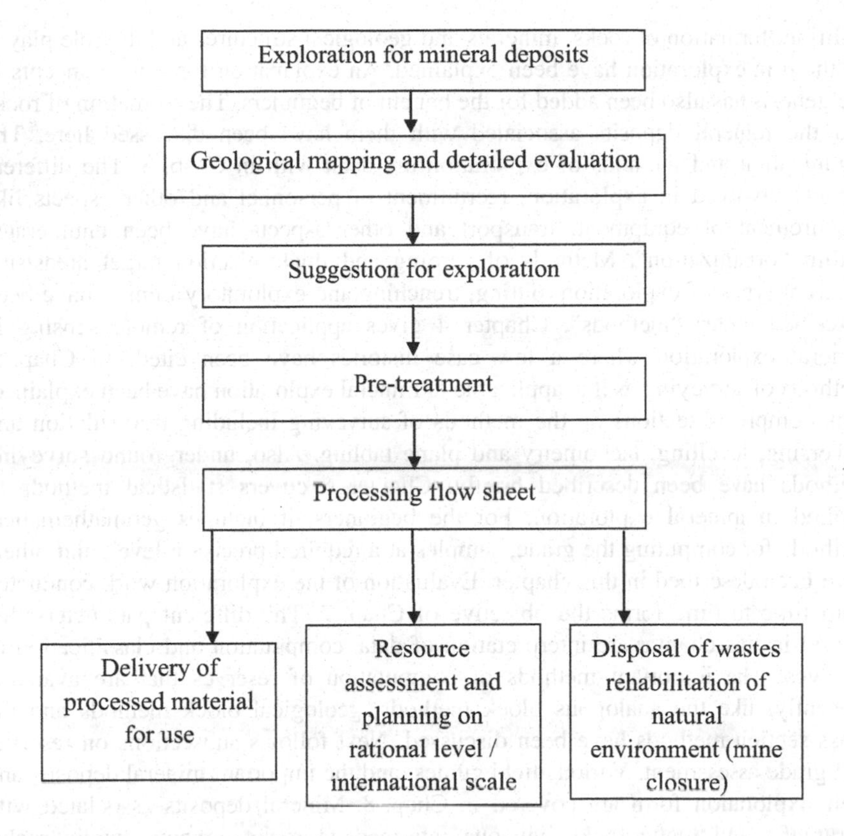

Fig. 1.4 Various aspects of exploration

time especially in a thickly populated country such as India. The prospector is giving place to specialized agencies and organizations. The prospector's simple tools are also getting progressively replaced by modern and sophisticated exploration instrumentation and highly trained specialized personnel. Mineral exploration today is a commercial activity demanding the use of many specialized skills. Mineral exploration all over the world has been spearheaded by earth scientists in general and geologists in particular.

This handbook aims serving as a guide and source of reference for the mining and exploration geologists, coal, oil and atomic minerals have been excluded from the purview of it. Among other minerals, only those which are currently under exploration and mining have been discussed. The handbook comprises chapters, includes several appendices and conversion tables. Each one of these chapters deals with different aspects of mineral exploration presented in such a way that the exploration geologist/and prospector could recapitulate the fundamentals before going into the more advanced aspects of the science of exploration geology (Chap. 1). Therefore, Chap. 2 discusses with rocks, minerals and geological structures and the role played by them in exploration with an explanation. In addition, formation of rocks, minerals and geological structures and the role played by them in exploration have been explained. An explanation for some concepts in ore genesis has also been added for the benefit of beginners. The formation of rocks and the mineral deposits associated with them have been discussed here. The organization and methods of exploration are dealt with in Chap. 3. The different aspects involved in exploration, recruitment of personnel and other aspects like procurement of equipment, transport and other aspects have been enumerated within "organization". Methods of zeroing and final selection target areas/site, different types of exploration, pitting, trenching and exploratory mining have been described under "methods". Chapter 4 gives application of remote sensing in mineral exploration where a few case histories have been cited. In Chap. 5, methods of surveying as it is applicable to mineral exploration have been explained. This comprises sections on the methods of surveying including triangulation and traversing, levelling, tachometry and plane tabling. Also, underground surveying methods have been described briefly. Chapter 6 covers statistical methods as applied in mineral exploration. For the beginners, it includes geomathematical methods for computing the grade, samples at a required precision level, and others have been described in this chapter. Evaluation of the exploration work conducted from time to time forms the objective of Chap. 7. The different parameters discussed in this chapter are interpretation of data, computation and classification of reserves. The important methods of computation of reserves that are available presently, like the analogous block methods, geological block methods and the cross-section methods have been discussed. Next follows subsections on reserves and grade assessment. Various field guides, and the important mineral deposits and their exploration form are covered in Chap. 8. Mineral deposits associated with ultramafic and mafic rocks, igneous intermediate rocks, igneous acidic rocks, sedimentary evaporate rocks, metamorphic rocks and products of residual weathering have been discussed separately. Field guides for the recognition of various mineral deposits have been summarized. In addition, guides to exploration of individual minerals have also been included in this chapter. Chapter 9 deals with exploration in producing mines. This includes sections on verification and correlation of earlier exploration data. The control of mining operations has also been described. Almost all aspects which have a direct bearing on mineral exploration find a place in handbook.

It needs to be understood that forecasts of the availability of raw materials have been made since the outset of the industrial age, but have proved to be a chain of miscalculation. Forty years ago, the prediction resulted in a wave of pessimism with regards to access to ore deposits. This possibly led to great exploration efforts in the 70s. However, the fear led to decrease in consumption during the recession of the 80s. This further led to surplus in supplies of ores and production capacity. It looks that there is no need to fear any shortage in the foreseeable future. Exploration and mining of ores is governed by consumption of metal. From the industrialized countries shift moved to emerging economies to meet expending industrial growth.

Let us understand that increasing world population and the efforts for higher standard of living shall lead to continuous rise in demand for metallic minerals. It looks that in the foreseeable future, the presently known supplies of ores shall be exhausted. It is thus imperative to search for new deposits.

Further Readings

Bailly PA (1978) Mineral exploration trends and prospects. Unpublished. Colorado School of Mines, Bolder, Co., USA
Charnley H (2003) Geosciences, environment and man. Elsevier, Amsterdam, p 527
Chaussier JB, Morer J (1987) Mineral prospecting manual. Elsevier, Amsterdam, p 300
Cox DP, Singer DA (1986) Mineral deposit models, vol 1693. USGS Bulletin, p 379
Eggert RG (1987) Metallic mineral exploration: an economic analysis, RFF Press, Washington DC, p 90
Evans AM (1999) Introduction to mineral exploration. Blackwell Science, p 396
Gandhi SM, Sarkar BC (2016) Essential of mineral exploration and evaluation. Elsevier, Amsterdam, p 410
Gass IG, Smith PJ, Wilson RCL (1979) Understanding the earth. ELBS, p 383
Haldar SK (2012) Mineral exploration—principles and applications. Elsevier, Amsterdam, p 372
Kursten M (1994) The importance of the geosciences to modern industrial society. Nat Res Dev 39:1–7
McLaron DJ, Skinner BJ (1987) Resources and world development. Wiley, New York p 940
Mitchell B (1979) Geography and resource analysis. Longman Scientific, Wiley, New York, p 386
Moon CJ, Whatelly MKG, Evans AN (2006) Introduction to mineral exploration and mining. Blackwell Publishing Professional, p 481
Pitchamathu DV (2011) Evolution of Indian mineral policy and its impact on mineral industry. Geological Society of India
Sinding-Larsan R, Wellmer FW (eds) (2012) Non-renewal resources—geoscientific and societal challenge. Springer, Berlin, p 215
Skinner BJ (ed) (1981) Economic geology special issue, 75th annual, p 964
Skinner BJ (1986) Earth resources, 3rd edn. Prentice Hall, p 184
Wellmer FW, Delheimer M, Wagner M (2007) Economic evaluation and exploration. Springer, Berlin

Chapter 2
Composition of the Earth and Mineral Resources

2.1 Crust of the Earth

The crust of the earth is about 30–70-km-thick form of the outer layer of the planet in which all rocks and minerals are found. Beneath the crystal layer is the zone known as the **Mohorovičić** discontinuity, which is followed by heavy rocks of a thickness of 3000 km. This is followed by a 1800-km-thick outer core of iron and nickel in a hot semi-plastic condition, which in turn is followed down to the centre by an inner core of 1000 km thickness.

The earth's surface of which the crust is a part is divided into the following components:

(i) lithosphere, which is the solid rock portion of the earth constituting the continents and extending beneath the ocean floor;
(ii) hydrosphere, which includes oceans, lakes and other water bodies, and rivers; and
(iii) atmosphere or the blanket of air which covers the continental crust.

The distribution of minerals in the earth's crust is shown in Table 2.1.

Out of 118 elements, as few as eight are known to be present in quantities exceeding one percent. The outer crust which is about 16 km deep is made up of the following elements—oxygen, silicon, hydrogen, calcium, sodium, potassium, magnesium, titanium, phosphorus, hydrogen, carbon and manganese. These elements constitute 99.5% of the crustal rock material. Elements like platinum, gold, silver, copper, lead, zinc, tin, nickel and others constitute the remaining 0.5%.

The hydrosphere, which comprises oceans, lakes, rivers and other water bodies, contains elements and compound in a dissolved form. The composition of water from various bodies is shown in Table 2.2.

As a whole, the crust contains broadly 1600 mineral species. Amongst these, 50 are rock-forming minerals of which 29 are most common. The remaining 1521 come from ore minerals. In order to understand the mode of occurrence of

G.S. Roonwal, *Mineral Exploration: Practical Application*, Springer Geology, DOI 10.1007/978-981-10-5604-8_2

Table 2.1 Major minerals in the crust of the earth

Mineral	Percentage of incidence
Feldspar	49
Quartz	21
Pyroxene, amphibole and olivine	15
Mica	8
Magnetite	3
Titanite and ilmenite	1
Others	3
	–
	100

Table 2.2 Composition of various water bodies (dissolved matter only)

Element/compound	Water of lakes, reservoirs of dams, smaller water bodies like rivers	Sea water
CO_3	35.15	0.41 (as HCO_3)
SO_4	12.14	7.68
Cl	5.68	55.04
NO_3	0.90	–
Ca	20.39	1.15
Mg	3.41	3.69
Na	5.79	30.62
K	2.12	1.10
SiO_2	11.67	–
Sr, M_3BO_3, Br	–	0.31
Total	97.25	100.00

economic mineral deposits, it is necessary to know about the various types of minerals and rocks, particularly their field occurrence, and methods of identification.

2.1.1 Common Rock Forming Minerals

A geologist defines a mineral as a naturally occurring, crystalline solid of specific chemical composition, structural arrangement of component atoms and physical properties, e.g. quartz, pyrite and diamond. Minerals combine to form rocks, defined and naturally occurring accumulation or mixtures of minerals formed by geological process, e.g. granite, limestone and oil shale. An inorganic substance naturally occurring in nature, though not always of inorganic origin, has (i) a definite chemical composition, more commonly with a characteristic range of chemical composition and (ii) distinctive physical properties or molecular structure. With a few exceptions, such as opal (amorphous) and mercury (liquid), minerals are crystalline and solids.

Although genesis is the major criterion for classifying rocks, a given rock derives its name, from characteristics and distinctiveness by virtue of its mineral constituent. In order to study the rocks, an understanding of minerals is a basic requirement. It may not always be possible to be conversant with all the known minerals. The best one may be taken for is to be able to recognize important minerals in the area of interest. Guide books for this are available to help in identifying the minerals in the field. Primarily aimed to guide basically mineral and ore collectors, they can also be used in exploration and prospecting.

2.1.2 Physical Characteristics of Minerals

There are certain common criteria to recognize minerals in the field. They are (i) colour, (ii) streak, (iii) lustre, (iv) hardness, (v) habit, (vi) fusibility, (vii) cleavage and fracture, (viii) tenacity and (ix) crystal system.

Form

This is one of the first observations made when a mineral is examined in a hand specimen. The form represents the common mode of occurrence of a mineral in nature. It is also called habit or structure of the mineral. To some extent, this is a function of the atomic structure of the mineral. Modes of formation, cleavage character, etc. of the mineral also contribute to the typical form of a mineral. Since some minerals consistently exhibit the same form, often it provides a valuable clue for mineral identification. But due caution should be exercised in arriving at a conclusion based on this property, because the same mineral may exhibit different forms or different minerals may exhibit the same form. So this should be treated only as a supporting clue, and the mineral should be identified by considering other clues also, obtained by different physical properties observed.

The following is the list of some common forms and the minerals (Table 2.3) which characteristically exhibits them, i.e. the appearance of a particular form is indicative of a certain specific mineral.

(i) Colour: The colour of a mineral is usually described in comparison with certain well-known objects of similar colour, e.g. ruby-red, leaf-green. Table 2.4 gives a summary of important characters of mineral.
(ii) Streak: Streak is the colour of the powdered mineral and is not influenced by impurities. It is identified by rubbing the mineral against a streak plate—a piece of white chert or any surface which can show the colour of the rubbed power. In the case of soft or powdery forms of minerals, the smear of powder on a piece of white paper will help its identification. The streak of certain important elements and minerals is given in Table 2.5.

Table 2.3 Characteristic forms of minerals

S. N.	Name of the form	Description	Example
1	Lamellar form	Mineral appears as thin separable layers	Different varieties of mica
2	Tabular form	Mineral appears as slabs of uniform thickness	Feldspars, gypsum
3	Fibrous form	Mineral appears to be made up of fine threads. Fibres may or may not be separable	Parallel fibres: asbestos types, satin spar *Radiating fibres*: Stibnite, some zeolites, malachite goethite, pyrite, pyrolusite
4	Pisolitic form	Mineral appears to be made up of small spherical grains (pea-size)	Bauxite
5	Oolitic form	Similar to pisolitic form but grains are of still smaller size (like fish eggs)	Some limestones
6	Rhombic form	Rhombic shape	Calcite, dolomite
7	Bladed form	Mineral appears as cluster or as independent lath-shaped (i.e. rectangular) grains	Kyanite
8	Granular form	Mineral appears to be made up of innumerable equidimensional grains of coarse or medium or fine size	Chromite, magnetite, pyrite
9	Reni form	Kidney-shaped. Mineral appears with number of overlapping smooth and somewhat large curved surfaces	Hematite
10	Botryoidal form	Similar to reniform but with smaller curved faces like bunch of grapes	Chalceodony, psilomelane, hematite
11	Mammillary form	Mineral appears with large mutually interfering spheroidal surfaces similar to reniform	Malachite
12	Acicular form	Mineral appears to be made up of thin needles	Natrolite, actinolite
13	Columnar form	Mineral appears as long slender prism	Tourmaline, precious topaz
14	Prismatic form	As elongated, independent crystals	Staurolite, beryl, apatite, quartz
15	Spongy form	Porous	Pyrolusite, bauxite
16	Banded form	Mineral occurs with numerous parallel bands (straight or curved) of similar or dissimilar colours	Agate
17	Crustal form	Polyhedral, geometrical shapes	Garnets, some zeolites, quartz, amethyst, pyrite, galena
18	Interpenetrating twin form	–	Staurolite, fluorite, pyrite, calcite
19	Massive form	No definite shape for mineral	Graphite, olivine quartz, jasper

(continued)

Table 2.3 (continued)

S. N.	Name of the form	Description	Example
20	Concretionary form	Porous and appears due to accretion of small irregularly shaped masses	Laterite
21	Nodular form	Irregularly shaped compact bodies with curved surfaces	Flint, limestone

Table 2.4 Characteristic colours of minerals

Colour	Mineral of element
Silver-white, tin-white	Native silver, antimony, arsenopyrite
Steel-grey	Platinum, manganite, chalcocite
Blue-grey	Molybdenite, galena
Lead-grey	Galena, stibnite
Iron-black	Graphite, magnetite, hematite
Black	Ilmenite, columbite, wolframite, mica, some amphiboles
Copper-red	Native copper
Bronze-red	Bornite, niccolite
Bronze-yellow	Pyrrhotite, pentlandite
Brass-yellow	Pyrrhotite, pentlandite
Gold-yellow	Chalcopyrite, millerite, pyrite
White with greenish tinge	Amphibole, pyroxene
Blue	Azurite, lapis-lazuli, sapphire, kyanite, beryl, fluorite, calamine
Green	Serpentine, malachite, spodumene, jadeite, talc, garnet
Yellow	Sulphur, orpiment, topaz, barite, sphalerite, siderite, goethite
Red	Ruby (corundum), garnet, cuprite, cinnabar, zircon, zincite, realgar, rhodochrosite
Brown	Staurolite, rutile, tourmaline, quartz

(iii) Lustre: Lustre is a measure of the reflectivity of the mineral surface and varies in degree and quality. Lustre is described as dull, feeble, brilliant and splendent. Other terms used to describe lustre are adamantine, resinous, pearl, vitreous, silky, metallic, etc., which are self-explanatory.

(iv) Hardness: Hardness is the resistance to abrasion and is expressed on the Moh's scale given in Tables 2.6 and 2.7 and a scale can also be used effectively in the field.

(v) Habit: Habit defines the size and shape of the crystal, and the structure and form taken by the aggregates. Crystals may be referred to as tabular (mica),

Table 2.5 Streaks of some important minerals

Streak colour	Element/minerals
Golden yellow	Gold
Silvery white	Silver
Copper-red	Copper
Greyish white	Platinum
Black	Pyrolusite, argentite, graphite (shining black), tetrahedrite, ilmenite, magnetite, columbite
Greenish black	Chalcopyrite, millerite, pyrite
Brownish black	Niccolite, pyrite, marcasite, wolframite
Greyish black	Chalcocite, bornite, galena, pyrrhotite, covallite (shining), stibnite, cobaltite, marcasite, arsenopyrite
Grey	Antimony, graphite (shining grey)
Brown	Sphalerite, tetrahedrite (Dark) rutile (pale)
Brownish red	Cuprite (shining), hematite, manganite
Brownish yellow	Goethite
Red	Cinnabar, pyrargyrite (dark), hematite
Orange-red	Realgar
Orange-yellow	Crocoite
Yellow	Orpiment (pale), vanadinite
Green	Malachite (pale), vivianite (very pale)
Blue	Azurite, lazurite
Purple	Vivianite

Table 2.6 Hardness scale of minerals

Mineral	Hardness on Moh's scale
Talc	1
Gypsum	2
Calcite	3
Fluorite	4
Apatite	5
Felspar	6
Quartz	7
Topaz	8
Corundum	9
Diamond	10

prismatic (quartz), or acicular (kyanite), depending on the ratio of length to thickness. Aggregates may be radiating (stibnite), fibrous (asbestos), bladed (mica), columnar (quartz), granular (hematite), etc. Minerals without any crystal form but showing circular outlines are described as coliform. Other terms in use are botryoidal (iron ore), reniform, etc.

(vi) Fusibility: This measures the ability of the minerals to melt. A scale showing the ease of the melting of various minerals is given in Table 2.8.

(vii) Cleavage and fracture: When a mineral is broken, it splits along a crystal place or an irregular surface. When the fracture occurs along a crystal plane, it is called cleavage and is described as indistinct, poor, good, perfect, or eminent. When the breaking is not along a regular plane, it is described as a fracture. Fracture is described as uneven, hackly, splintery, fibrous, earthy, or conchoidal. Some common types of fractures shown by the most typical minerals are given in Table 2.9 and tenacity in Table 2.10 which measures the ease with which mineral breaks are described below for some minerals.

For the criteria described above area guide, several criteria are required for a correct identification of a mineral properly. There are, however, some ores and minerals which can be identified by a single criterion.

Table 2.7 Determination of hardness during exploration work

Hardness		
2–2.5	Can be scratched by	Fingernail
3–3.5	Can be scratched by	Copper coin
5.5	Can be scratched by	Glass, knife blade
6	Can be scratched by	Vesuvianite
6.5	Can be scratched by	Vesuvianite
7	Can be scratched by	file

Table 2.8 Scale of fusibility

Mineral	Melting point, °C
Stibnite	525
Natrolite (chalcopyrite)	965
Almandite garnet	1200
Orthoclase	1200
Actinolite	1296
Bronzite	1380
Quartz	1600

Table 2.9 Types of mineral fracture

Fracture	Mineral
Conchoidal	Obsidian, flint, chalcocite, sphalerite, quartz, halite
Subconchoidal	Rutile, stibnite, argentite, cordierite, staurolite
Even	Galena
Uneven	Cinnabar, millerite, chalcopyrite
Hackly	Native iron

Table 2.10 Tenacity of common minerals

Tenacity	Mineral
Brittle	Calcite
Sectile	Gypsum
Malleable	Native gold and silver
Flexible	Talc

2.1.3 Importance of Mineralogical Characteristics

2.1.3.1 Grain Size

The classification of any material according to the size of occurrence or the size to which they need to be crushed or pulverized is very important and, therefore, the sizes have to be described with precision.

The particle sizes of some of the naturally occurring material are given in Table 2.11.

In the case of pulverized material, there are standard sieve sizes as per the British standard, the ASTM standard, or the Indian standard. All of these specifications in terms of mesh sizes are not always common in all. Therefore, it is essential to know in inches or in millimetres the exact specifications and standard tables to the same (see Table 2.12).

The grain sizes are of great significance in the beneficiation tests. Normally, the microscopic size varies from 0.001 to −0.25 mm, whereas the megascopic sizes range from +0.25 to 100 mm or 150 mm which could be identified by magnifying lens. The effect of grain size has a great bearing on the method of beneficiation. The best results under various methods are expected at the following grain sizes.

Table 2.11 Size ranges of naturally occurring materials

Main classification	Secondary classification	Diameter in mm	Example
Gravel	Boulder	256	Conglomerate
Gravel	Cobble	256–64	Conglomerate
Gravel	Pebble	64–4	Conglomerate-
Gravel	Granule	4–2	Conglomerate
Sand	Very coarse	2–1	Sandstone
Sand	Coarse	1–0.5	Sandstone
Sand	Medium	0.5–0.25	Sandstone
Sand	Fine	0.25–0.125	Sandstone
Sand	Very fine	0.125–0.062	Sandstone
Silt	–	0.062–0.005	Siltstone
Clay	–	<0.005	Shale

Table 2.12 Size separation and minerals

	Method of beneficiation	Best result at grain size, in mm
1	Heavy media up to specific gravity of the medium (water, salt bath, etc.) being 5.2	2.0–10
2	Heavy media solutions	10.0–80.0
3	Jigging	1.0–25.0
4	Tabling	0.1–2.5
5	Humphrey spiral	0.1–1.0
6	Cyclone	0.5–3.0
7	Magnetic concentration	0.075–75.0
8	Electrostatic separation	0.10–2.5
9	Froth floatation	0.01–0.25

Table 2.12 emphasizes the importance of the grain size of the mineral to which the ore has to be ground for making it suitable for beneficiation by various methods.

2.1.4 Mineralogical Identification

Mineralogical identification is a necessary need in the initial stages of prospecting and exploration but more so in the adoption of specific process of mineral beneficiation and processing whenever such beneficiation is called for. To cite an example, the general term bauxite may cover both monohydrate and trihydrate ore minerals either as boehmite and diaspore or gibbsite. The mineralogy of bauxite and the loss on ignition as per the chemical composition are stated to be related. An exploration geologist is required to give special attention to texture and to the ore mineralogy of bauxite since it would guide in the production of alumina.

Similarly, in the case of chromites, though they are spinels with the general composition of $FeCr_2O_4$ some are rich in iron and in some others magnesium replaces iron. A combination of precise chemical and mineralogical analysis shall guide if chromite ore meets a grade with the desired Cr/Fe ratio.

In the case of iron ores, the tolerance of FeO which is the index of magnetite content (Fe_3O_4) in the otherwise hematite (Fe_2O_3) ore is important, and the presence of hydroxide minerals such as limonite and goethite is considered deleterious. Magnetic separation can be worked on magnetite iron ores. Likewise, if the deposit is mainly limonite and goethite ores, it will be possible to work them as direct pelletizing ores by driving out the LOI and enriching the ores.

Similarly, we would like to know if zinc ores are of gahnite (aluminate) or sphalerite (sulphide), lead ores are of galena (sulphide) or cerussite (carbonate) and titanium ores are of rutile or ilmenite minerals.

Nature of mineralogy is important for the above examples in respect of ore minerals need to be kept in mind. In fact, an understanding of ore mineralogy is pre-requisite to plan mineral processing. A basic knowledge will be of help for exploration geologists as to the ore processing since understanding of ore textures will help in deciding the appropriate method and grinding of ores for the best release of metals.

2.1.4.1 Specific Gravity of Minerals

The specific gravity of a mineral is one of the important criteria for its identification. Application of specific gravity in exploration and subsequently a proper evaluation of tonnage, ore processing and beneficiation are significant.

The in situ bulk density (true specific gravity) is the sum total of the weighted average specific gravities of all the mineral present, weighted by the proportion in which the different minerals are present. Therefore, except while dealing with a mono-mineral deposit without any gangue or associated minerals, which is rare, in all other cases, the bulk density has to be determined. Besides, the word bulk density connotes specific gravity corrected for porosity voids, joints, etc. The best practical approach is to determine how much a specific volume of the material weighs and then to determine the volume-to-weight ratio directly. For example, if an accurately measured one cubic metre (m^3) of ore in situ is taken out and weighs 3 tonnes, the bulk density of the material would be 3.

To determine the specific gravity of dense and non-porous rocks like igneous and metamorphic rocks, a Walker's Steelyard balance is used. In other cases such as bauxite and sandstone, which are porous, the specific gravity is determined by applying wax on the specimen and using a steelyard balance. However, a pycnometer is used for determining the true specific gravity of porous material. It is possible to use geophysical methods, such as densitometer tool, for determining the bulk density that is measuring the bulk density using radioactive method.

In multi-metal ores having highly variable specific gravity of ore minerals such as chalcopyrite and galena (4.2 and 7.5), it would be essential to find out the proportion of the minerals present in the ore and then apply the respective specific gravities for the two ore minerals and a common specific gravity for all the gangue minerals so that the bulk density for the ore as a whole can represent a weighted average. Unless duly weighted to the proportion of minerals of varying specific gravities, the tonnage factor adopted will not be accurate.

It will be useful to use or prepare ready-made tables (such as Table 2.13) so that an idea of what tonnage factor is to be considered for ores of different specific gravities could be ascertained. Table 2.13 shows the tonnage factor to be considered for various specific gravities at one percent base metal assay. This table can be conveniently used for varying percentages of assay of a combination of assays.

Table 2.13 Tonnage factors for specific gravity

S. No.	Specific gravity	Tonnes of ore per m^3	kg. of base metal in m^3
1	2.5	2.5	25
2	3.0	3.0	30
3	3.5	3.5	35
4	4.0	4.0	40
5	4.5	4.5	45
6	5.0	5.0	50
7	5.5	5.5	55
8	6.0	6.0	60
9	6.5	6.5	65
10	7.0	7.0	70
11	7.5	7.5	75

2.2 Formation of Rocks and Mineral Deposits

In Geological evolution of the lithosphere, there are three main processes responsible for the formation of rocks and mineral deposits. They are (i) magmatism, (ii) sedimentation and (iii) metamorphism. These three processes give rise to three major groups of rocks, respectively, viz.

(1) igneous rocks, (2) sedimentary rocks and (3) metamorphic rocks

Since most of the mineral deposits are associated genetically with these three types of rocks, a study for understanding of these rocks is required and helpful in mineral exploration programme.

2.2.1 Igneous Rocks

All igneous rocks are formed from cooling of molten magma. Varying conditions of genesis, coupled with varying chemical and mineralogical composition of the magmas, have been instrumental in producing various types of rocks. Igneous rocks have been classified on the basis of crystallization/grain size, mineral and chemical composition. A field classification is shown in Table 2.14 based on texture, mineral content and chemical composition of individual rocks is ideally suited for the needs of exploration geologists.

Table 2.14 Igneous rocks—criteria for recognizing in the field

(i)	Dimensions	Shape and structural relationship to adjacent rocks
(ii)	Contacts	Sharp, transitional, shape (plane, undulating, grooved, irregular with dimensions of irregularities), structure (jointing, faulting, brecciation) metamorphism (width of zone, mineralization, texture attitude strike, dip, etc.)
(iii)	Colour	Wet, dry, fresh, weathered conditions
(iv)	Composition	Minerals recognizable with hand lens and their estimated proportions, estimated composition and proportion of ground mass
(v)	Texture	Degree of crystallization, porphyrite, equigranular, etc.
(vi)	Structure	Platy or linear with attitudes, columnar jointing, flow perlitic, spherulitic, orbicular, gneissic or other
(vii)	Hardness	Friability, partings
(viii)	Erosion	Erosion and weathering products
(ix)	Identification	Interpretation as to the mode of emplacement, e.g. hydrothermal or other

2.2.1.1 Guide to Study and Recording Igneous Rocks in the Field

In studying igneous rocks, it is important to follow a set routine, so that their identification becomes easy, particularly in the field. Broadly, igneous rocks may be either (a) intrusive or (b) extrusive.

(a) Intrusive igneous rocks: the following sequence of observations, with the accompanying terminology, may be used in studying and reporting intrusive igneous rocks.
(b) Extrusive igneous rocks: In studying and reporting extrusive igneous rocks, the following sequence may be employed.
(i) Dimensions—Width, length and thickness; (ii) Shape or variation in dimensions; (iii) Relation to overlying or underlying adjacent formations; (iv) Contacts, top and bottom described pre-existing surface, textural difference, fault, overlap, attitude and alteration effects; and (v) Type of accumulation—Pyroclastic or flow, viscous liquid breccia flow and reworking by wind or water.

By sequential recording of data under the headings discussed above and systematic correlation and analysis, full or partial identification of most of the igneous rocks is possible. Where precise determinations are required, thin sections need to be studied. The aim of an exploration geologist is to locate ores and minerals which form economic deposits. An exploration geologist often looks for any changes that have occurred in the lithology such as hydrothermal alteration, weathering and mineral occurrences.

2.2.1.2 Guide to Identification of Igneous Rocks in the Field

Table 2.14 classifies the igneous rocks rather broadly. There are several other rocks with intermediary compositions. Identification of rocks in the field is very important. Some of the criteria which help in identifying the major rock units in the field are given below:

(a) **Granite**: Minerals—feldspar, quartz, with minor minerals like biotite, hornblende, magnetite, etc. Minerals are easily seen without the help of a hand lens. Minerals may be intergrown. The rock is light in colour, and moderate specific gravity means moderate weight. Granite is generally hard and tough. Common economic mineral deposits are cassiterite, wolframite, galena, sphalerite and others coming through hydrothermal fluids, and gold coming in epithermal quartz veins.
(b) **Pegmatite**: Minerals—feldspar, quartz, muscovite and very coarse grains. Minerals show intergrowth texture. Pegmatites are light in colour as well as weight. They occur in a vein-like pattern. The economic mineral deposits associated with pegmatites are mica, quartz, feldspar, tourmaline, beryl and a host of others.
(c) **Granite porphyry**: This rock has a granite ground mass, phenocrysts of felspar, quartz, etc.
(d) **Monzonite**: Minerals—feldspar, biotite, hornblende and pyroxene. Minerals are mutually intergrown and are visible to the naked eye. The rock is light coloured and light in weight. The economic mineral deposits associated with granite porphyry and monzonite are copper, lead, zinc, gold, etc.
(e) **Syenite**: Minerals—feldspar, biotite and hornblende. They are visible to the naked eye and are intergrown. The rock is light in colour and weight. Sometimes they form porphyries with phenocrysts of feldspars. When syenite contains nepheline mineral, it is called nepheline syenite. Corundum deposits show a genetic relationship with nepheline syenite.
(f) **Diorite**: Minerals—amphiboles, biotite or pyroxenes, plagioclase and feldspars. This rock is grey or dull green in colour.
(g) **Gabbro**: Minerals—feldspar, pyroxene, hornblende and olivine. Mutually intergrown minerals are visible to the naked eye. In colour, the rock is dark and of heavy weight. Gabbro and anorthosite show economic concentrations of titaniferous magnetite, magnetite and ilmentie.
(h) **Periodotite**: Minerals are composed of olivine, pyroxene and hornblende, and are visible to the naked eye and exhibit mutual intergrowth. The rock is usually dark in colour and heavy in weight. Peridotites occur as intrusives. Diamonds occur in kimberlite, a variety of peridotite. Other deposits are of platinum and chromite.
(i) **Basalt**: Minerals composed of pyroxene and olivine and are not visible to the naked eye. The rock is fine grained and dark grey to black in colour and cannot be split into layers. It occurs as lava flows, dykes and sills.

2.2.1.3 Mineral Deposits Associated with Igneous Rocks

During the course of emplacement of magma, various minor and accessory constituents start getting progressively concentrated till a stage is reached when the end product is rich in some constituent which under favourable conditions gives rise to economic mineral deposits. These favourable conditions are generated by three processes:

(a) magmatic concentration
(b) contact metasomatism
(c) hydrothermal processes.

Magmatic Concentration

Magmatic concentration occurs as a result of cooling/crystallization, or concentration by differentiation of intrusive igneous rocks. Deposits of this type are associated with intermediate and deep-seated intrusive igneous rocks. Two stages of magmatic concentration have been recognized, e.g. (i) early magmatic and (ii) late magmatic. The important processes are dissemination, segregation and injection. Dissemination is a simple crystallization process without any concentration. Diamond bearing kimberlite is formed by this process. Segregation is the process of crystallization, differentiation and accumulation. Chromite and corundum deposits are formed by this way.

Contact Metasomatism

During the consolidation of a magma, high-temperature gases emanate from it. These gases usually contain certain mineral matter in gaseous form. When these gases travel through various rocks which are already existing, metamorphism and metasomatism (high-temperature replacement) take place. Certain rocks are particularly amenable to these chemical reactions and such favourable rocks become a preferred target for the accumulation of economic mineral deposits. The effect of this reaction is to chemically transform the already existing minerals into new forms, such as limestone and dolomite become marble, carbonaceous matter becomes graphite and sandstone becomes quartzite.

All magmas do not give rise to conditions of contact metamorphism. It is exclusively associated with intrusive magmas, which give rise to intrusive bodies like stocks and batholiths. Rocks which give rise to contact metamorphism generally have a granular ground mass which suggests a slow cooling of the magma. Rocks which are most susceptible to contact metasomatism are limestone, dolomite and limestone with impurities like silica, alumina, iron, manganese and others. Any rock structures, such as cleavage, bedding plane, joints and fracture systems, accelerate the invading process.

Mineral deposits associated with contact metamorphism are relatively small in size and irregular in shape, and therefore difficult to locate in the field. An understanding of field occurrence is micro-tectonics and structures of rocks. Since deposits of this type occur near intrusive bodies, granular ground mass, as seen in impure calcareous rocks, helps in locating them. The most typical field evidences for locating contact metamorphism deposits are chilled or quenched borders, evidence of dolomitization, effects of baking, hardening, partial or full recrystallization near contacts.

Economic mineral deposits are formed when the contact metamorphism takes place due to intrusions of quartz monzonite, monzonite, granodiorite, diorite and others.

The common deposits resulting are iron, copper, zinc, lead, tin, tungsten, molybdenum, emery, garnets and corundum.

Hydrothermal Process

The end product of the emplacement of magma is a fluid which generally carries metals in solution. These liquids get injected into the country rocks which offer the maximum pore spaces and other openings like fracture and fault planes. By the processes of cavity filling and replacement, various mineral deposits of economic importance are formed. Various stages of mineralization are recognized in the hydrothermal process. The three stages most commonly recognized are (i) hypothermal, (ii) mesothermal and (iii) epithermal. These stages have been recognized on the basis of certain distinct temperature and pressure conditions accompanying the formation of minerals.

Hydrothermal deposits are formed under certain optimal conditions such as (i) the availability of mineralizing solutions capable of dissolving and transporting mineral matter, (ii) the availability of openings in the rocks through which the solutions may be channelized, (iii) the availability of sites for the deposition of mineral content, (iv) chemical reactions promoting the formation of deposits and (v) sufficiently concentrated mineral matter to form a workable deposit.

These processes give rise to the following types of deposits: (i) fissure veins, (ii) shear zone deposits, (iii) stock works, (iv) saddle reefs, (v) ladder veins, (vi) pitches and flats, (viii) breccia fillings, (viii) solution cavity fillings, (ix) pore space fillings and (x) vesicular fillings.

In hydrothermal deposits also, there is no distinct ore to rock associations which could help recognize the deposits in the field. However, the combination of shear zones and intrusive igneous ore bodies nearby provides indications for ore search and locating a deposit. Typical field guides for locating hydrothermal mineralization are alteration haloes, sericitization, argillic alteration, silicification, chloritization and serpentinization which can readily be recognized as they are clearly visible on the rocks and thus on the ground.

2.2.2 *Sedimentary Rocks*

The sedimentary or detrital rocks are those formed by the deposition of solid materials carried in suspension by the agencies of transport. A number of different sedimentary rock units can be recognized.

Weathering of rocks produces soluble and insoluble components: The soluble products, like calcium and magnesium, are carried away in solution. The solids are transported by water, wind, etc. Deposition of the soluble components takes place when the carrying solutions reach the point of chemical saturation usually in bodies of water which are relatively calm. The process of precipitation may be augmented by the presence of bacteria. Iron and manganese are considered to have deposited by this process. However, the most common deposits are limestone and dolomite.

The solids wear down to small sizes: When the transporting medium loses its natural velocity, deposition takes place. The resulting rocks may be sandstone, shale, siltstone, etc. Mineral deposits of economic value are found in some of these rock types, but without any known genetic link. However, placer deposits are by-products of the process of sedimentation.

Table 2.15 shows general classification of sedimentary rocks. For more precise classification, the reader may refer to standard books on the subject.

Table 2.15 Classification of sedimentary rocks

Rock type	Inorganic	Organic
Talus	Coarse fragmentary material resulting from weathering	Lime made from shells—chalk, coral rock, etc.
Breccia	The above when cemented	Silica from the shells of plants diatomaceous earth, etc.
Soil	Unsorted material resulting from rock weathering	Carbon from plants, peat, lignite, coal, etc.
Gravel	Coarse fragments rounded by the action of water and wind	Hydrocarbons from animals—petroleum asphalt, amber, etc.
Conglomerate	The above when cemented	Phosphates from animals—Guano, phosphate rocks, etc.
Sand	Finer material deposited by water or wind	
Sandstone	The same material when cemented	
Clay	The finest material mostly kaolin, deposited by water	
Loess	The finest material deposited by wind	
Shale	The same material when cemented	
Marl	Fine particles of lime, pure or impure	
Limestone	The same material when cemented	
Till	Unsorted material left by glacial ice	
Tillite	The same material when cemented	

Methods to study sedimentary rocks in the field: In studying and reporting sedimentary rocks in the field, the following sequence and terminology may be adopted:

(a) External form of rock unit: lenticular persistent, very regular in thickness, dimensions, relation to overlying or underlying units.
(b) Colour: Colour of the unit as a whole, wet or dry colour of individual particles.
(c) Bedding:

  (i) How manifest, sharp by parting, by difference in texture, colour, transitional.
  (ii) Shape of bedding surface, plane, undulating, ripple marks, irregular, if not plane; record details of form and dimensions of features.
  (iii) Thickness of beds: comparative thickness and different orders. Relation of thicknesses, rhythmic, random, etc, and if variable, relation between thickness and composition and bedding.
  (iv) Attitude and direction of bedding surface: horizontal, inclined and curved. Relation to each other; parallel, intersecting, tangential; angles between different attitudes and directions, dips, strikes, dimensions, relation of size, composition, shape to attitude and direction; and relation of composition to different types of bedding.
  (v) Markings of bedding surface: mud cracks, rain prints, bubble impressing, ice crystal impressions, trails and footprints.
  (vi) Disturbances of bedding: edge-wise of intraformational conglomerates, folding or crumbling of individual beds before consolidation.

2.2.2.1 Guide to Study Sedimentary Rocks in the Field

The more common sedimentary rocks are (a) sandstone, (b) limestone, (c) shale and (d) conglomerate. They are identified in the field as follows:

(a) **Sandstone**: Mineral—quartz, cemented together by silica, lime or iron oxide. Coarse sandstone grades into conglomerate whereas fine sandstone grades into sandy shale. Sandstones may show fossil, current bedding ripple marks.
(b) **Limestone**: Mineral—calcite; individual grains are invisible to the naked eye, a stony appearance. Effervescence is given in the HCl acid test.
(c) **Shale**: Mineral—clay; individual grains are invisible to the naked eye, a stony appearance; it can be split into layers.
(d) **Conglomerate**: Minerals—quartz, feldspars and various other rocks and mineral pieces; individual minerals can be easily identified with the naked eye. The grains have normally varied sizes and shapes. Sand fillings or some other matrix can be seen between the grains.

There are also different intermediary varieties of specific type within sedimentary rocks.

2.2.2.2 Mineral Deposits Associated with Sedimentary Rocks

Two major ore-forming processes are involved in this. They are (a) sedimentation and (b) evaporation. Besides, the processes of sedimentation are manifested in the formation of placer and certain residual deposits. These will not, however, be considered here.

Sedimentation

Sedimentation involves various processes. Detritus material gathered from various sources by water are transported to the site of accumulation, usually a basin, where the material is deposited. Compaction, diagenesis and chemical alteration follow, giving rise to mineral deposits.

All rocks contain elements like iron, copper, manganese, etc. in varying proportions and in combination with other compounds. These elements are released from their parent rock during weathering.

Solutions charged with carbonic acid, humic or other organic acids react on the rocks to dissolve elements. Certain minerals like clay which are not chemically active are carried away in suspension. The solutions containing such suspension remain stable so long as there is no change in their physical and chemical environment. As soon as there are some changes in the above conditions, the dissolved elements precipitate and are deposited on the floor of the basin. There are also bodies of water which are fed by mineralizing solutions like fumeroles and hot springs. A stage is reached when the dissolved minerals precipitate to form deposits.

Evaporation

By a process of rapid evaporation, the dissolved material can accumulate to form deposits. Such deposition is typical in arid and desert terrains.

Important mineral deposits formed by sedimentation include iron ore, manganese ore, copper ore and uranium ore, and industrial minerals like limestone, phosphorite, gypsum, salt and clay.

2.2.3 *Metamorphic Rocks*

Metamorphic rocks are formed by the mineralogical and structural adjustment of solid rocks to new physical or chemical conditions that happen at depths below the surface zones of weathering and concentration and which differ from the conditions under which the rocks in question are originated. They are classified on the basis of

(a) field occurrence, (b) structure and texture, (c) mineralogical composition and (d) chemical composition .

Four major metamorphic processes are recognized.

(a) cataclastic metamorphism,
(b) thermal metamorphism,
(c) dynamo-thermal metamorphism and
(d) plutonic metamorphism

(i) **Cataclastic metamorphism**: Here, the minerals are crushed and granulated through the development of small amounts of stress and low temperatures. The typical features produced by this process are crush breccias, micro-breccias, mylonites, flow cleavage, fracture cleavage and stain–slip cleavage, which guides in recognizing the phenomenon in the field.
(ii) **Thermal metamorphism**: This is connected to the intrusion of large igneous rock bodies. Metamorphic changes take place due to the heat of intrusion and produce zones of mineral and textural around the intrusive, which is recognizing an aureole. The typical rocks produced by this process are hornfels, calc-silicate hornfel, quartz-hornfel, crystalline limestone, marble, serpentinous rocks, etc. Mineral deposits of economic value are asbestos, limestone, marble and graphite.
(iii) **Dynanomothermal metamorphism**: In this process, the rocks are recrystallized and ions formed by directed pressure and heat. The rock produced are phyllite, mica schist, quartz schist and gneisses. Economic mineral deposits are soapstone, talc, sillimanite, kyanite and andalusite.
(iv) **Plutonic metamorphism**: Changes take place in rocks due to the combined effect of great heat and uniform pressure, a condition of great depths. Typical products are granulites, leptites, leptynites and gneisses.

Method to study metamorphic rocks in the field: The following sequence and terminology shall help to study metamosphric rocks in the field (Table 2.16).

Table 2.16 Terminology used in describing metamorphic rocks in the field

(a)	Type of metamorphism	Cataclastic, thermal, dynamo-thermal or plutonic
(b)	Form and field name of the rock unit	Shape, lenticularity, regularity of thickness and shape, dimensions, etc.
(c)	Structural relation to adjacent formulation	
(d)	Contacts	(i) How manifest; sharp, transitional intrusive (ii) Shape of contacts: plane, undulating, grooved, irregular, record of dimensions (iii) Strike and dip (iv) Disturbances of contacts: intraformational conglomerates, brecciation, jointing, faulting, alteration

(continued)

Table 2.16 (continued)

(e)	Colour	Colour of the mass as a whole; wet or dry, colour of individual parts on particle, inclusions
(f)	Composition	List of identifiable minerals and proportion of each; compositional banding, inclusions, lateral or vertical variations
(g)	Texture and structure	(i) Degree of crystallization and granularity, porphyroblasts, relic phenocrysts or pebbles (ii) Foliation, gneissic, schistose, slaty, banded, lenticular (iii) Contortions of compositional bands or foliation (iv) Relic textures and structure, ripple marks, spherulites, flow lines (v) Lateral or vertical variations in texture and structure
(h)	Hardness	Friability, flakiness, cases or parting due to foliation
(i)	Erosion and weathering products	Soil, sediments

2.2.3.1 Guide to Recognizing Metamorphic Rocks in the Field

Some criteria which help identify a few of the metamorphic rocks in the field are given in Table 2.17.

Table 2.17 Field identification of metamorphic rocks

(a)	Hornfels	Non-schistose rock of equidimensional grains. Occur typically in contact aureoles
(b)	Buchites	Partially-fused hornfelsic rocks occurring as xenoliths in basalts, diabases, etc.
(c)	Slates	Fine-grained rocks with perfect planar schistosity but lacking in segregation banding
(d)	Phyllites	Similar to salates, but grains are coarser. New mica and chlorite impart a lustrous sheen to schistosity
(e)	Schists	Strongly schistose, commonly lineated metamorphic rocks in which the grains are coarse enough to allow microscopic identification of the component minerals
(f)	Gneisses	Coarse-grained irregularly banded rocks with discontinuous rather poorly defined schistosity. They are products of high-grade regional metamorphism
(g)	Granulites	Even grained metamosphic rocks, poor in mica and rich in quartz, feldspar, pyroxenes and garnet which lack a prismatic or tabular habit
(h)	Mylonites	Fine-grained, flinty-looking, strongly coherent, banded or streaked rocks resulting from extreme granulation of coarse-grained rocks without any special chemical composition
(i)	Cataclasites	These are rocks formed by ruptural deformation. Cataclasites may grade into mylonites

(continued)

Table 2.17 (continued)

(j)	Phyllonites	These rocks resemble phyllites, but are formed by mechanical degradation of initially coarse rocks
(k)	Quartzites	Metamorphic rocks composed of recrystallized quartz. Quartzites are generally produced by regional metamorphism of sandstones
(l)	Marbles	Marbles are produced by the regional metamorphism of calcareous sediments. The rock is composed of calcite or dolomite
(m)	Amphibolites	Metamorphic rocks composed or hornblende or plagioclase
(n)	Serpentines and soapstones	Composed of serpentinous minerals, talc, chlorite, etc. and formed by metasomatism of peridotites

2.2.3.2 Mineral Deposits Associated with Metamorphic Rocks

The process of metamorphism may alter an old deposit and produce a new one or may act on any rock and produce deposit, provided the original minerals are conductive to such a transformation. The source rocks undergo recrystallization or recombination or both. The major deposits produced by metamorphism are asbestos, graphite, talc, soapstone, sillimanite, kyanite and garnet. Mineral deposits formed due to metamorphism are in the form of whole rocks like marble or in lenticular, linear concentrations like soapstone and asbestos. These deposits can form from any favourable source rock and hence do not generally show any rock to ore deposit association or give any clear field evidence for their location.

2.3 Other Ore-Forming Processes and the Resultant Mineral Deposits

Apart from those discussed above, there are two other processes which have given rise to important mineral deposits. These are:

(a) Mechanical and residual concentrations and
(b) Oxidation and supergene enrichment.

(a) **Mechanical and Residual Concentration**

Due to the continuous action of weathering agents, rocks are disintegrated mechanically and decomposed chemically. Unstable minerals like feldspars, pyroxenes, amphiboles, etc. are chemically altered, and the compounds are dissolved, olivine transported by water and wind. Stable minerals like quartz and gold are not transformed chemically but released out of enclosing matrix. Due to the continuing action of transporting agents like running water and wind, the particles get worn down to very small sizes and are transported to great distances. Deep and continuous weathering offers a large variety of products in the form of mechanical fragments and in the dissolved chemical form. The action of weathering and transportation creates two types of mineral deposits:

(i) Residual concentrations and
(ii) Mechanical concentrations (placer formation).

(i) **Residual Concentration**: Due to weathering and transportation, various rock constituents are removed; the residues accumulate till they attain sufficient concentration, purity and size to form a mineral deposit. Certain conditions are necessary for their formation, such as a rock containing valuable minerals, favourable climate, conditions of chemical decay and a mode of selective transportation where only the undesirable constituents are washed off. Mineral deposits which have formed by this process include iron ore, manganese, bauxite, clay, nickel, phosphate, barites, tin and ochre.

(ii) **Mechanical Concentration**: Mechanical concentration is basically a physical separation of the lighter constituents from heavier constituents accomplished by running water or moving air. This takes place in two stages, viz. (i) weathering and separation of stable minerals from their matrix and (ii) their concentration. The minerals involved in this process may come from the already existing mineral deposits or from rocks which contain some valuable mineral deposits or from rocks which contain some valuable mineral constituents in a much disseminated form. The resultants are known called deposits which may be of four types, viz. (a) eluvial placers, (b) stream or alluvial placers, (c) beach placers and (d) eolian placers.

(a) **Eluvial Placers**: Eluvial placers are formed on hill slopes. Material released from outcrops upslope is roughly sorted, the heavier staying close to the outcrop and the lighter moving downhill. Some field guides to locate such deposits are areas of breaks in the slope, hillside talus, scree accumulations, etc. Important deposits of tin, gold iron and manganese are formed by this way.

(b) **Alluvial placers (stream placers)**: Minerals and rocks released during weathering are transported downstream by rivers and streams. During floods, the material is carried rapidly downstream. Whenever there is a fall in the velocity of water, the heavy minerals settle down. A sufficient concentration of one mineral ultimately gives rise to an important mineral deposit. Such deposits are formed in meander bends and near natural obstructions in the stream course. Placer deposits of gold and diamond are formed in this way. Some field guides to locate such deposits are meander bends, stream junctions, alluvial fans, cones, accumulation near points of a sudden drop in velocity, etc.

(c) **Beach Placer**: Due to wave and shore action, placers are formed along seashores. Sorting of heavy and light minerals takes place due to wave action. Deposits formed this way include gold, ilmenite, magnetite, monazite and diamonds. Some field guides are meander bends, unusual colouration in a beach sand, sparkle and scintillation effects in reflected sunlight, etc.

Table 2.18 Important products of the oxidation of ore deposits

Metal	Original composition	Oxidized product
Iron	Sulphides Carbonates Oxide	Hematite, limonite sulphate Limonite, ferric hydroxide Hydrous ferric oxide
Copper	Sulphides	Carbonates, oxides, native copper, silicate
Zinc	Sulphides	Carbonate, silicate
Lead	Sulphide	Sulphate, carbonate
Tin	Oxide or sulphide	Oxide
Aluminium	Silicate	Oxide, silicate

(d) **Eolian placers**: These are formed in acid regions. The material is released during weathering in desert or arid by wind action and sorted during transportation. When the wind current meets with an obstruction, the heavy minerals settle down. Economic deposits of salt and gypsum found in Rajasthan desert are formed in this way.

(b) **Oxidiation and Supergene Enrichment**

Mineral deposits get exposed to the atmospheric action as a result of weathering. Surfaces water results in the outcrops yielding solvents which in turn dissolve another one. This process takes place up to the top of the water table. If the solutions penetrate the water table, their metallic content gets precipitated and rich secondary cross develops. Both oxidation and enrichment have produced base metals deposits.

Important products of oxidation of primary ore deposits are listed in Table 2.18.

2.3.1 *Guide to Locate Oxidation Enrichment Ores*

Look for the presence of gossans cover, zone of leaching and limonitic caps. Gossans have a high diagnostic value for the buried mineral deposits.

Gossan capping is a result of chemical reaction taking place within the rocks and mineral as a sequel to interaction/reaction of water and air. This occurs near the surface or very near the surface. The surface features thus produced help in guiding towards mineralization. Gossan represents a combination of oxides, silicates, carbonates and finally sulphides. Sulphides enrichment is a chemical process, explained as the secondary environment. Physical feature of oxidized caps is an indicator of sulphides, as concealed ore body. Gossans represent a high level of weathering but retain distinctive characters which help in locating mineralization. In southeast Rajasthan at Dariba-Rajpura, India, Cu, Zn–Pb mineralization was

Table 2.19 Gossan caps and expected buried sulphide mineralization

	Nature of gossan	Nature of deposit
(a)	Form and size	Generally the outcrop faithfully outlines the shape of orebody
(b)	Collapsed gossan, voids in gossan	
	(i) Abundant	Sulphides
	(ii) Shape	If square, galena, pyrite
	(iii) No void	No important deposit likely
(c)	Colour of limonite	
	(i) Brown, maroon, orange, etc.	Copper
	(ii) Yellow, brick red	Pyrite
	(iii) Deep brown/brick red, yellowish	Chalocopyrite
	(iv) Chocolate	Bornite
	(v) Deep maroon	Chalcocite
	(vi) Tan to brown	Sphalerite
	(vii) Orange	Galena
(d)	of box work	
	(i) Coarse, cellular with blebs, masses,	Chalcopyrite
	coarse and angular walls	Bornite–chalcopyrite
	(ii) Fine, cellular, thin, small, friable walls	Sphalerite
	spaces, blebs	Sphalerite
	(iii) Coarse, cellular, siliceous, thin, rigid	Sphalerite
	angular walls	Bornite
	(iv) Cellular spongy	Chalcocite, covellite, bornite
	(v) Fine cellular, shriveled	Chalcocite bornite
	(vi) Triangular, crusted, curved	Chalcocite
	(vii) Porous	Galena
	(viii) Pitchlike limonite, no cells	Galena
	(ix) Limonite crusts	Galena
	(x) Cleavage	Molybdenite
	(xi) Diamond mesh	
	(xii) Pyramidal	
	(xiii) Foliated	

discovered through gossan. However, it is important to make a distinction between a gossan and a false gossan. An example of false gossan is also available in NW Rajasthan in Khetri copper belt. While Cu-mineralization exists at Khetri, the southern side of this plunging anticlinorium does show gossan but are only pyrites bearing, as seen at Saladipura. Broadly, we know that gossan related to economic mineralization show anomalous geochemistry as compared to the hydrous, iron oxides as seen in Saladipura in NW Rajasthan, India. Some of the diagnostic features of gossans are listed in Table 2.19.

Table 2.20 Zoningin Gossan cover

(i)	Presence of vertical zoning	Oxide—top followed by supergene sulphide enrichment and primary—protore
(ii)	Presence of gossans and capping	As described above
(iii)	Mineralogy	Sooty chalcocite, covellite, native silver, native gold, marcasite for sulphides and goethite, hematite for iron ore, pyrolusite and psilomelane for manganese

Generally, the process of oxidation gives rise to secondary enrichment which produce rich sulphide deposits below the zone of oxidation. Criteria for the recognition of such enrichments are listed in Table 2.20.

2.4 Geological Structures

It has been discussed earlier that mineral deposits are formed under complex geological environments. Although the formation of a deposit is largely controlled by the ore-forming processes, it is the geological structure which helps in localizing mineral deposits. These structures may be regional or purely local. Our knowledge of the structure of earth's crust is derived from the continents. The continents themselves offer two regions which have a separate tectonic and structural history, the mobile belts and the cratons. The mobile belts are those characterized by igneous activity and carthquakes. There are also areas of rapid sediment accumulation giving rise to geosynclines and geanticlines. When a mobile belt is characterized by geosynclines and geanticlines they are called orogens which are the loci of mountain building activity thus structurally complex. Compared with the mobile belts, the cratons are static and stable. From the point of view of ore-genesis, the structurally complex mobile belts of the geological past are important mineral deposit sites. A study of the elementary to the complex type of geological structures is a part of exploration geology. Structures thus studied are as follows:

(a) Structure of igneous rocks,
(b) Structure of sedimentary rocks and
(c) Structure of metamorphic rocks.

2.4.1 *Structure of Igneous Rocks*

Two types of rock structures are recognized in igneous rocks: (a) structure due to flow and (b) structure due to fracture.

(a) **Structure due to flow**: Two types of structures may be identified.

(i) Linear flow structure: The parallel orientation of needle-shaped inclusions constitutes linear flow structure.

(ii) Platy flow structure: The parallelism of the flat surfaces of tabular, or platy inclusions like phenocrysts, xenoliths and schlieren (flow layer) constitutes platy flow structures. It may lie in the plane of foliation but may form an angle with it.

(b) **Structures due to fracture**: Under this come joints, sheets and faults.

(i) Columnar jointing: The rock is divided into hexagonal column formed at right angles to the cooling surface.

(ii) Joints: Joints which lie perpendicular to the flow lines are termed cross joints (Q joints) and tension joints are formed due to the upwelling of the liquid magma in the centre of the intrusion; 'S' joints are steeply dipping joints which strike parallel to the flow lines.

(iii) Sheeting: They consist of gently curved joints which divide the rock into flat lenses parallel to the topographic surfaces. When they are closely spaced, they are called mural joints.

(iv) Faults: Normal and thrust faults of purely local significance are seen on the border of large intrusive. The extent of slip along the planes is usually very small. A number of such faults may be arranged in echelon. Flat lying normal faults are also not uncommon.

Besides these, broad regional structures like ring dykes and cone sheets are also present.

2.4.2 *Structures of Sedimentary Rocks*

A large number of structures are recognizable in sedimentary rocks. Some of them are described below:

(a) **Beddings or stratification planes**: Sedimentary rocks are arranged in layers. The plane which separates the various layers is called bedding or stratification plane.

(b) **Graded bedding**: Beds show some gradation in size from bottom to top. The coarser grains are at the bottom and the finer ones are at the top. This textural arrangement of sedimentary bed is known as graded bedding.

(c) **Initial dip**: Initial dip is different from dip exhibited in an exposure. Initial dip is the slope of the stratification plane during sedimentation. It is common for sediments formed in basins to exhibit this dip.

(d) **Discordant bedding**: Normally bedding planes and beds are parallel to each other. When this parallelism is lost, the bedding planes become discordant.

Terms like current bedding, cross bedding and false bedding are also applied to discordant bedding.

(e) **Ripple marks**: Ripple marks are those ridge like prominences seen on sediments. These are created by the movement of air or water over unconsolidated sediments.

(f) **Rain, Drip and Hail Impressions**: These are impressions formed on loose sediments by rain and hails, dripping from trees or plants. The upper surface of these is concave which can help in recognizing the orientation of beds.

(g) **Mud cracks**: Mud cracks are formed by the exposure of soft mud to sun's rays. They have wide mouths and tapering bottoms which help in recognizing the top and bottom of beds.

Various other structures are also noticeable in sedimentary rocks. Since they are of no direct use in studying the strata sequence (top and bottom of beds particularly), they are not being discussed. The structures described above can be directly used for establishing the top and bottom of sedimentary formations.

2.4.3 *Structures of Metamorphic Rocks*

No distinct structures which are exclusive to metamorphic rocks can be recognized unless features like schistosity and gneissosity are considered. Although they are primary to metamosphic rocks, they are modifications of existing structures of the original rocks. Interpretation of rock structures is very important in studying major and minor geological structures. In certain types of geotechnical studies connected with open pit design and stability of slopes also, the rock structures are important.

2.4.4 *Other Structures*

2.4.4.1 Dip and Strike

When strata are affected by tectonic forces, structures developed, which tells the position orientation of rocks in place. This comprises two factors known as strike and dip. Strike refers to the direction in which a geological structure (such as a bed, a fault plane, or a joint plane) is present. When an inclined bed is suitably exposed on the surface, its direction of occurrence, its direction of inclination and amount of inclination can be actually measured directly by a clinometer. The strike direction is defined as the direction of the trace of the intersection between the bedding plane and a horizontal plane. The dip amount is the angle of inclination between the bedding plane and a horizontal plane. The dip shows the direction along which the inclination of the bedding plane occurs. As regards inclined strata, in a direction perpendicular to the strike direction, the inclination of bedding planes is maximum

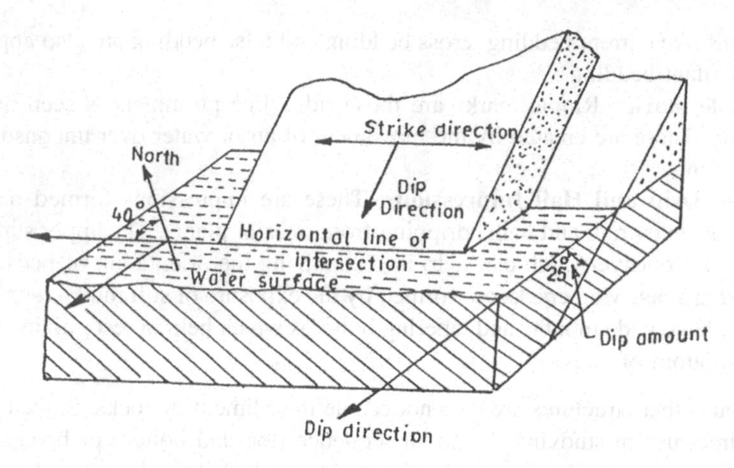

Fig. 2.1 Strike and dip relationship

and these are called the true dip direction or simply dip direction. The amount of inclination along this direction is called the true dip amount or simply dip amount Fig. 2.2. Close observation reveals that this dip amount gradually decreases on either side of the true dip direction towards the strike direction. Along the strike direction the inclination of lie bed (i.e. dip amount) is zero, i.e. the bedding plane will be horizontal. All directions which lie in between the strike direction and the true dip direction are known as apparent dip directions and the inclinations along them are called apparent dip amounts. Apparent dip amount is higher towards the true dip direction and lesser towards the strike direction. The strike direction and true dip directions are always mutually perpendicular (Fig. 2.1).

In the field, exploration geologist uses clinometer compass and the Brunton compass for the purpose.

2.4.4.2 Folds

When a set of horizontal layers are subjected to compressive forces, they bend either upwards or downwards. The bends noticed in rocks are called folds. Folds are described variously as wavy, arch-like, curved, undulating, or warping appearances found in rocks. They are also called flexures or buckling phenomenon of rocks. Generally, they occur in series. In nature, folds found in rocks have a range of magnitude in terms of their length and breadth. At one extreme, they may be very small such as few cm in length. At the other extreme, they may be a few km across and several km long. The radius of curvature of a fold is small compared to its wavelength and amplitude. In terms of their nature too, folds may occur as single local bend (monoclines) or may occur repeatedly and intricately folded according to the tectonic history of the region.

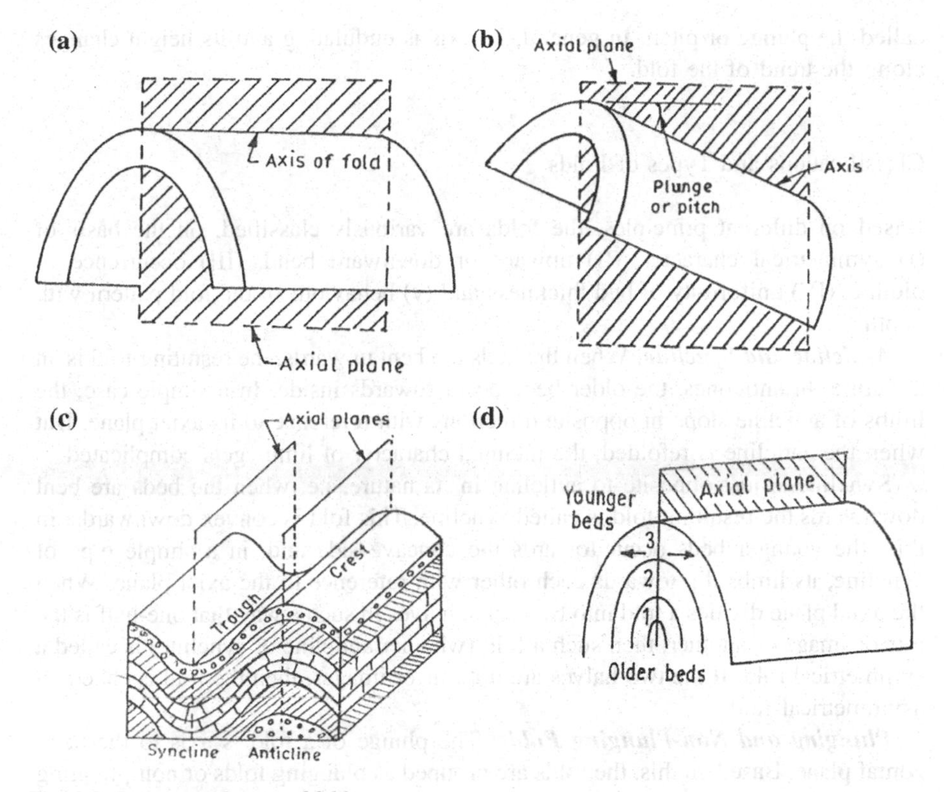

Fig. 2.2 Patterns and parts of folds

Limbs, crest, trough, axial plane, axis, wavelength and plunge, which are the important parts of a fold, are shown in Fig. 2.2a–d.

Limbs or Flanks

These are the sides of a fold. There are two limbs for every fold and one limb common to the adjacent folds as shown in Fig. 2.2.

Crest and Trough: The curved portions of the fold at the top and bottom are called crest and trough, respectively. In general, these are smoothly bent, but in chevron folds, these are sharp and angular. Some geologists refer to crests and troughs as hinges.

Axial Plane and Axis: This is an imaginary plane which divides a fold into two equal (or nearly equal) halves. It passes through either the crest or the trough. Depending upon the nature of the fold, the axial plane may be vertical, horizontal or inclined. In case of symmetrical folds, the axial plane divides the fold into exactly two equal halves. But in asymmetrical folds, the two halves will be only nearly equal. Like a bedding plane, the axial plane also can be expressed by strike and dip refers to the trace of the intersection between the axial plane and the crest or trough of the fold. Depending on the nature of the fold, it may be inclined or horizontal or vertical. When it is inclined, the angle between the axis and the horizontal plane is called the plunge or pitch. In general, an axis is undulating and its height changes along the trend of the fold.

Classifications and Types of Folds

Based on different principles, the folds are variously classified, on the basis of (I) symmetrical character, (II) upward or downward bend, (III) occurrence of plunge, (IV) uniformity of bed thickness and (v) behaviour of the fold pattern with depth.

Anticline and Syncline: When the beds are bent upwards, the resulting fold is an anticline. In anticlines, the older beds occur towards inside. In a simple case, the limbs of anticline slope in opposite directions with reference to its axial plane. But when the anticline is refolded, the inclined character of limbs gets complicated.

Syncline is just opposite to anticline in its nature, i.e. when the beds are bent downwards the resulting fold is called syncline. This fold is convex downwards. In this, the younger beds occur towards the concave side and, in a simple type of syncline, its limbs dip towards each other with reference to the axial plane. When the axial plane divides a fold into two equal halves in such a way that one-half is the mirror image of another, then such a fold (whether anticline or syncline) is called a symmetrical fold. If the two halves are not mirror images, then the fold is called an asymmetrical fold.

Plunging and Non-Plunging Folds: The plunge of a fold is axis to the horizontal plane. Based on this, the folds are grouped as plunging folds or non-plunging folds. In geological maps, when strike lines are drawn for both the limbs, for a

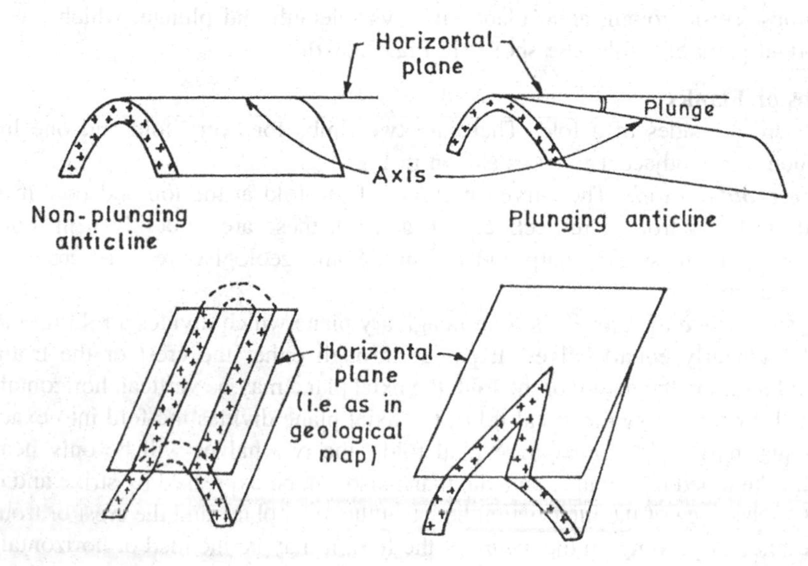

Fig. 2.3 Plunging and non-plunging anticlines

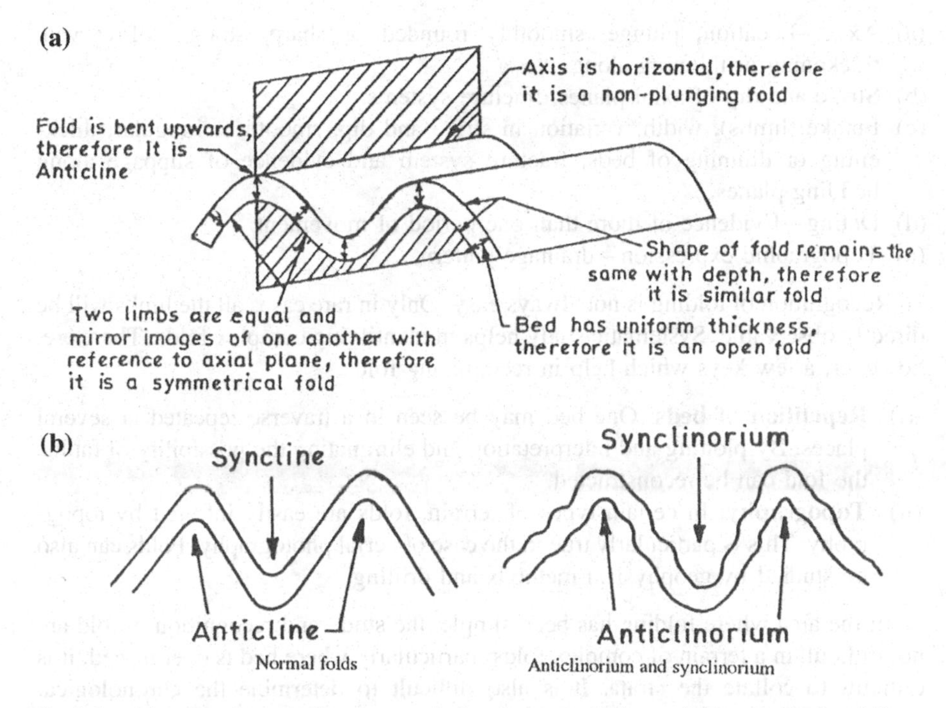

Fig. 2.4 **a** Non-plunging, similar, symmetrical and open anticline and; **b** concept of anticlinorium–synclinorium

non-plunging fold, they will be mutually parallel and for a plunging fold, they will be either converging or diverging but not parallel. On a horizontal plane, the outcrops of non-plunging and plunging folds appear as shown in Fig. 2.3. Reader needs to refer to books on structural geology for details.

Geanticlines and geosynclines: The anticlines and synclines with a normal shape but a very large magnitude are called "geanticlines" (giant anticlines) and "geosynclines". Geosynclines are of special importance from the physiography point of view because these are the places accompanied by long periods of sedimentation and are the potential places which can become great mountain ranges in future such as the Himalayas, the Alps, the Appalachian and the Cordillera, a few examples sites of mountain ranges. When the limbs of folds are not plain but characterized by the appearance of other minor folds on them, the major folds are called anticlinoriums and synclinoriums (Fig. 2.4a, b).

Criteria for Recognition of Folded Strates

In order to study folds systematically to enable easy interpretation, the following sequence of observation and terminology is recommended:

(a) Axes—Location, plunge, smoothly rounded or sharp, straight of curved, thickening or thinning along axes.
(b) Strike and dip of axial planes, fracture systems.
(c) Flanke (limbs), width, variation in strike and dip, smooth or irregular, thickening or thinning of beds, fracture system and evidence of slippage along bedding planes.
(d) Dating—Evidence of more than one period of movement.
(e) Topographic expression—drainage pattern.

Recognition of folding is not always easy. Only in rare cases all the limbs will be directly observable. Systematic study helps in identifying complex folds. There are, however, a few keys which help in recognizing folds.

(i) **Repetition of beds**: One bed may be seen in a traverse repeated at several places. By plotting and interpretation and eliminating the possibility of faults, the fold can be reconstructed.
(ii) **Topography**: In certain types of terrain, folds are easily inferred by topography. This is particularly true in the case of aerial photography. Folds can also be studied by geophysical methods and drilling.

In the area where folding has been simple, the study and recognition of fold are not difficult in a terrain of complex folds; particularly where bed is over turned, it is difficult to collate the strata. It is also difficult to determine the chronological sequence of beds. Some criteria for recognizing the top and bottom of the bed are essential. A few criteria adopted are discussed below.

When beds are overturned, primary features are used in recognizing the top and bottom. These are cross bedding, ripple marks, graded bedding and mud cracks. In rocks, the vesicular tops and flow structures can be used for correct orientation of such minor features to be confirmed by field observation. In ripple marks, the crests are at the top. In cross bedding, the laminae are parallel to the bedding at the bottom, but form sharp angles at the top. In graded bedding coarser grains are always at the bottom.

2.4.4.3 Faults

Structurally, faults fractures along with relative (i.e. parallel) displacement of adjacent blocks have taken place. If such relative displacement does not take place on either side of fracture plane, it is called a joint. Thus both joints and faults are fractures in rocks but with a difference in the kind of displacement. Joints may be described as a set of aligned parallel cracks or openings in geological formations.

Magnitude and Nature of the Faulty Plane: Faults also have considerable range in their magnitude. Some occur for short lengths (a few centimetres), while others can be traced for very long distances (to several kilometres). The degree of displacement differs widely. In some, the displacement may be less than a centimetre while in others it may be of many metres or even kilometres (as in the case of nappes of the Himalayas). The magnitude of faulting depends on the intensity and the nature of shearing stresses (kinds of tectonic forces) involved. The relative displacement caused during faulting may be horizontal, vertical or inclined, and may be parallel or rotational with reference to the fault plane.

Rarely, the displacement during faulting occurs along a single fault plane. In many cases, faulting takes place along a number of parallel fractures. That is the total displacement is distributed over this zone. Such a zone which contains a number of closely spaced subparallel fractures along which the relative displacement has been taken place is called the shear zone or fault zone: Due to frictional resistance, the brittle rock masses get crushed to different degrees in the fault zone (producing fault breccias). Sometimes, this crushing may produce an even fine clay-like material called "gouge".

Fault Plane: This is the plane along which the adjacent blocks were relatively displaced. It is the fracture surface on either side of which the rocks had moved past one another. Its intersection with the horizontal plane gives the strike direction of the fault. The direction along which the fault plane has the maximum slope (i.e. the direction which is perpendicular to the strike direction) is its true dip direction. The amount of inclination of the fault plane with reference to the horizontal plane along the true dip direction is called its (true) dip amount. Sometimes, the term "hade" is used to refer to the angle between the inclined fault plane and a vertical plane. Naturally, both dip and hade together make up 90°.

The fault plane may be plain and straight or may be curved or even irregular. It may be horizontal, inclined or vertical. Like a bedding plane, a fault plane too is described by its attitude by its strike direction, dip direction and dip amount.

Parts of a Fault

The different parts of a fault are shown in Fig. 2.5.

Foot Wall and Hanging Wall: When the fault plane is inclined as shown in the figure, the faulted block which lies below the fault plane is called the "footwall" and the other block which rests above the fault plane is called the "hanging wall". In the case of vertical faults, naturally, the faulted blocks cannot be described as footwall or banging wall.

Slip: The displacement that occurs during faulting is called the slip. The total displacement is known as the net slip. This may be along the strike direction (strike slip) or the dip direction (dip slip) or along both as shown in Fig. 2.5.

Heave and Throw: The horizontal component of displacement is called "heave" and the vertical component of displacement is called "throw". In the figure, the points A and B were together side by side before faulting. After faulting, they are displaced and hence found separately in different positions. In this, AC which is the horizontal component of displacement is heave and CB which is the vertical component of displacement is throw of a fault (Fig. 2.6). In vertical faults, there is only throw, but no heave. In horizontal faults, there is only heave, but no throw.

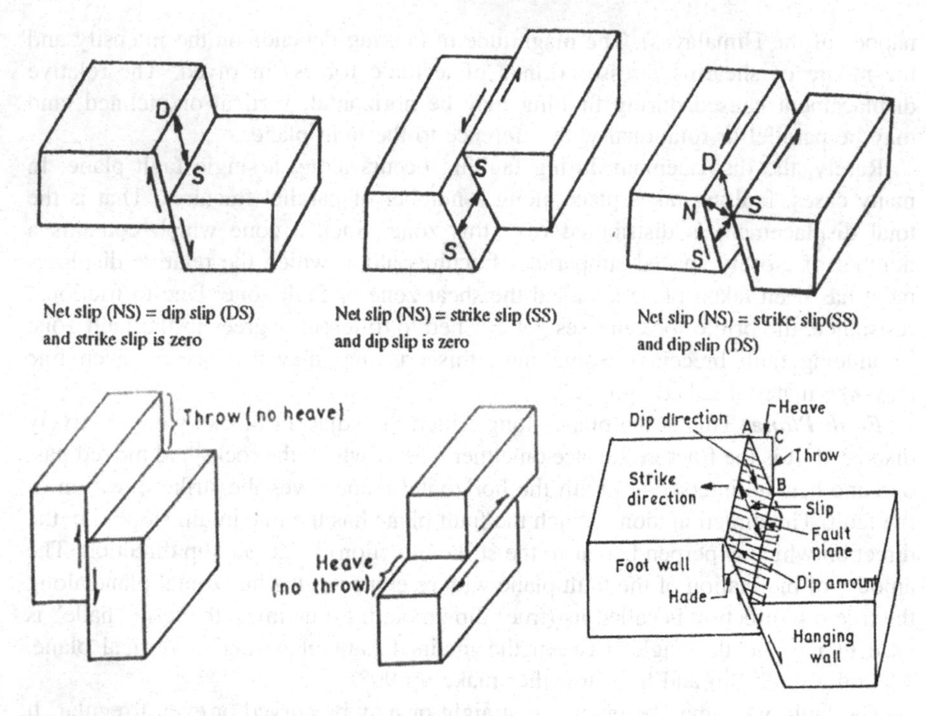

Fig. 2.5 Understanding of parts of a fault

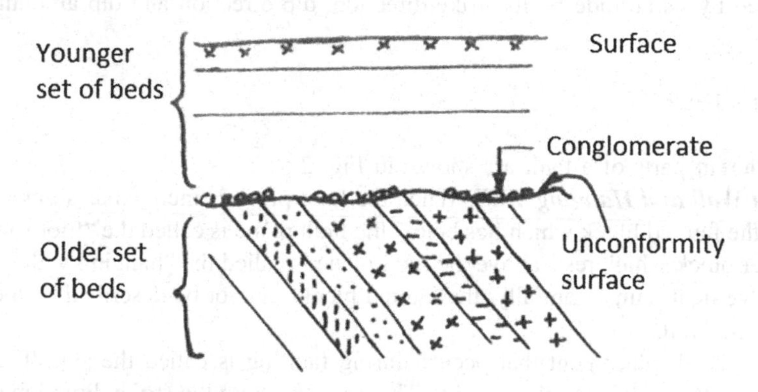

Fig. 2.6 Parts of an unconformity

Absolute and Relative Displacement: Faults are studied only after they have occurred and hence the actual or absolute type of displacement suffered by rocks (i.e. which faulted block had actually gone up or gone down and to what extent) is not known. The relative position of blocks as seen in the fields is possible through five different kinds of relative displacements as shown in Fig. 2.6. In view of the absence of knowledge of absolute movements during faulting, the figures of faulting are shown with symbols representing relative displacement (half arrows).

Recognition of Faults During Exploration Field Work

In several cases faults are recognizable, but when difficult, careful observation of certain features of faults is essential for deciphering the fault. Six criteria are as follows:

(a) Discontinuity of structures: Features like dykes, sills, veins, prominent fractures and folds may be seen ending abruptly against a place in an exposure. This may be due to faulting.
(b) Repetition and omission of strata: In a traverse line, strata may disappear altogether in a sharp contact or may be repeated sequentially in association with sharp contacts. Both may indicate faulting.
(c) Features characteristic of fault planes: In some cases, it is possible to recognize the fault plane by virtue of characteristic features such as (i) slickensides, (ii) mullion structure, (iii) drag, (iv) gauge, (v) breccias, (vi) mylonite and (vii) horses (caught up blocks).
    (i) Slickensides: Striations in the fault planes caused by movement and can be easily recognized by the glistening surfaces and striations.
    (ii) Mullion structures: Large grooves or furrows with a definite crest and bottom.
    (iii) Drag: The end of the beds affected by fault is dragged up and down.
    (iv) Gouge: Fine-grained clay-like powdery rock.
    (v) Breccia: Mixture of angular and subangular rock pieces of varying sizes in a finely crushed matrix.
    (vi) Mylonite: Microbreccias with streaked or platy structure, which is typically dark and fine grained; the coherence of a microbreccia is maintained during deformations.
    (vii) Hores: Small blocks of rocks caught up in a wide fault plane.
(d) Silicification and mineralization: Silicification may occur along the zones of fracture. Similarly, mineralization may also occur.
(e) Sudden change in the sedimentary facies: The phenomenon of a coarse-grained sandstone abutting against a shale of the same age and other similar instances is indicative of faulting during sedimentation.
(f) Physiographic evidences: The following physiographic evidences are suggestive of faults: (i) offset ridges, (ii) scarps, (iii) triangular facet and (iv) truncation of structures by a ridge front.

(i) Offset ridge: Where a resistant sedimentary stratum may show discontinuity. This is called an offset ridge; (ii) Scarp: A steep straight slope of any height; (iii) Triangular facet: On scarp faces, some 'V' notches may form due to erosion during movement. The sum total is a structure known as a triangular facet; (iv) Truncation of the structure by ridge front: Sudden termination of structures, particularly against a mountain front, is suggestive of faulting.

Other field evidences are springs in a linear arrangement, such as trees in a line and lakes in a line, all of which may coincide with the alignment of the fault. The sudden steep in a stream bed, abrupt ending of a stream, also may indicate faulting.

The criteria recommended here are indicative of faulting. Besides, some of them also indicate other structures like folds, unconformity. Careful study and a series of eliminations at each stage are needed before a particular set of criteria can be correlated to a fault. Usually, a combination of criteria is applied before any conclusion is drawn.

Common Difficulty Related to Faults in Mineral Exploration

When a tectonic disturbance takes place in an area, it may give rise to new structures either by the partial or total destruction of old structures or by superimposition of a new set of rocks over the relief of existing structures. Faults, folds, shears and joint are geological deformation caused by stress and strain. In a normal sequence of events, folds are formed first, whether synclines or anticlines; and if the forces causing the folding are still intense, then the rocks yield to these forces forming faults and shears. In an area which is geologically un-disturbed, it is possible to establish the sequence of events. But, in an area that has experienced disturbances belongs to repeatedly, it is difficult to establish which generation an individual structure to and its effect on other structures.

The effect of folds on faults is widely understood. However, it can be attributed that the effect of faults on folds is still a subject of discussion. Many examples have been given cited to show the influence of faults on folds, though on a regional scale.

Just like folds, faults also present problems to the exploration geologist. Some fault planes act as passages for the mineralizing solutions, and the deposits are formed in the fault plane itself. Faults may cut off mineralizing solutions in a favourable host rock or even act as a connection to a favourable host rock. Faults may block off the already existing deposits. As in folds, in faults also, it is important to establish the exact relationship existing between faults and mineralizing solutions. If the fault is pre-mineral, every change in the fault plane is influencing the mineralization. On the other hand, if the fault is post-mineral, the deposit may be merely displaced. The following criteria are sometimes useful in distinguishing the pre-mineral faults from the post-mineralization faults:

(a) Pre-mineralization faults:

- (i) Mineralization will be on the fault plane. Ore may be in breccia and vugs and
- (ii) Localizing effect of the fault will be on ore. Ore bodies tend to be in the fault.

(b) Post-mineralization fault:

(i) Ore will be slicken sided or brecciated. Drags will be clearly visible in many cases and
(ii) Observable offsetting of veins or orebody.

The presence or absence of a fault in a region influences an exploration strategy particularly in the selection of methods. Ore bodies may come close to the surface in certain faults making it an attractive target proved easily by shallow drill holes. It can also affect the economic workability of an ore body adversely by throwing it down to great depths.

In some cases, the determination of down-thrown or up-thrown block is critical. The criteria for recognition are discussed below:

(i) Correlation of wall rocks—If the sequence of the rocks which have been faulted is known, it is easy to find out the displaced block.
(ii) Drag—As explained earlier, the drag is always against the direction of movement.
(iii) Slicken sides—The groove of a slickenside will be smooth in the direction of movement, and rough against it.
(iv) Throw of minor and sympathetic fault—Since minor and sympathetic faults form in harmony with the main faults, their direction of throw will indicate the direction of the throw of the main fault.

2.4.4.4 Joints

As already mentioned under faults, joints are fractures found in all types of rocks. They are cracks or openings formed due to various reasons. Naturally, the presence of joints divides the rock into parts or blocks. These separated blocks may move only perpendicular to the plane of fracture, but do not move past one another as in the case of faults. Though the joints may be described as cracks in rocks, they may differ mutually. The difference between them is somewhat similar to that of fracture and cleavage (both of which refer to the nature of the broken surface) found in minerals. Joints, like cleavage of minerals, occur oriented in a definite direction and as a set (as a number of parallel planes). The cracks, like the fracture of minerals, are random or irregular in their mode of occurrence. Thus, joints occur, generally, parallel and oriented fractures in rocks. Such a group of fractures is called "joint set". Every set of joints (i.e. joint planes) shall have their own strike and dip. As mentioned earlier, the folded and faulted sedimentary rocks when associated with tension and shearing joints, the limestones and sandstones are characterized by a set of parallel fractures which enables these rocks to be split into thin sheets of uniform thickness. These rocks are called flaggy sandstones and flaggy limestones. In some sedimentary rocks, two sets of joints which are mutually perpendicular (and parallel to strike and dip directions) occur at right angles to the bedding planes. Joints are common as parallel to the bedding planes of sediments; in sandstones and limestones, joints occur several metres apart but in shales they occur closely.

Joints in Metamorphic Rocks: Though joints are found very commonly in metamorphic rocks also, they do not have any definite pattern of occurrence in any specific kind of rocks.

In this context, it is appropriate to know a little about rock cleavage or fracture cleavage. When shear between beds occurs either due to folding or faulting, a complimentary system of fractures often develops. In the weak beds, these fractures are generally closely spaced. Because of these fractures, the rock can split into thin sheets. Such a phenomenon is called rock cleavage or fracture cleavage.

2.4.4.5 Unconformity

Unconformity is one of the common geological structures found in rocks. It is somewhat different from other structures like folds, faults and joints in which the rocks are distorted, deformed or dislocated at a particular place. Still, unconformity is a product of diastrophism and involves tectonic activity in the form of upliftment and subsidence of land mass.

When sedimentary rocks are formed continuously or regularly (at a stretch) one after another without any major break, they are said to be a set of conformable beds, and this phenomenon is called conformity. All beds belonging to a conformable set shall possess the same strike direction, dip direction and dip amount. On the other hand, if a major break occurs in sedimentation (if sedimentation does not take place for a long interval) in between two sets of conformable beds, it is called an unconformity. So, an unconformity refers to a period of nondeposition and appears as a plane of contact between two sets of conformable beds.

There are different types of unconformities. Figure 2.6 shows a very common kind of unconformity and its parts. Generally, other unconformities through similar may differ in detail. All types have two sets of formations belonging to two different ages, i.e. one set is older and the other set younger having a depositional break in between. They have an unconformity surface; a few may have conglomerates along unconformity surfaces.

Formation of an Unconformity

The formation of an unconformity involves three stages as follows: In (i) the first stage, under favourable conditions, a conformable set of beds is formed; (ii) in the second stage, a break in sedimentation (deposition of sediments stops) occurs. This happens generally due to the upliftment of that region to the level above that of the water level (of river, lake, where deposition had been occurring). After some time, when subsidence occurs in that region and favourable conditions for sedimentation recur, (iii) the third stage begins. Here, renewed sedimentation produces another set of conformable beds. The set of beds, i.e. those which have been formed during the third stage, is younger compared to the other set of beds formed earlier (i.e. those which have been formed during the first stage).

The aforementioned indicates that an unconformity is a plane (or a slightly undulating surface) which represents a break in sedimentation. An unconformity also represents a period of nondeposition of sediments and separates two sets of conformable beds which belong to different ages.

(a) **Hiatus**

An unconformity which represents a long geological period (during which break in sedimentation had occurred) is known as a "hiatus". In India, examples of such a hiatus are seen in the Peninsular region, where, after the formation of Precambrian strata (i.e. the Cuddapah system and the Vindhyan system), there was no deposition of sediments for a long geological time (this time gap covers the Cambrian period of 150 million years, the Ordovician period of 90 million years, the Silurian period of 35 million years, the Devonian period of 50 million years and a part of the Carboniferrous period which is 60 million years, i.e. total approximately 385 million years). After that, the sedimentation began the Gondwana group of sediments that were deposited.

(b) **Conglomerates and Bauxite along an Unconformity**

The two important points about unconformity formation are as follows: (i) When the first set of beds is uplifted and a break in sedimentation takes place, geological action (weathering and erosion followed by transportation and deposition) may or may not occur during the unconformity period. If it occurs, the exposed beds are disintegrated and the rock fragments roll and get rounded or subrounded during their transport. Later, such pebbles and gravels get deposited under favourable conditions. When sedimentation resumes (younger series of beds begin to form), the pebbles of older beds get cemented and a conglomerate is formed. Thus, conglomerates occur in association with some unconformities.

During the unconformity period, if considerable leaching of earlier rocks occurs, these rocks may get changed over to laterites or bauxites. Thus, bauxite deposits occur sometimes along unconformities. (ii) The second important point is that the first formed conformable set of beds mayor may not get tilted (or folded) either during their upliftment or subsidence, depending upon the way the relevant tectonic forces had acted. If they were in a tilted disposition at the time of formation of the second (i.e. younger) set of beds, then beds of both these sets will be mutually inclined; otherwise, they will be mutually parallel.

Types of Unconformity

Based on factors such as the types of rocks, relative attitudes of sets involved and their extent of occurrence, the different types of unconformities are named.

Nonconformity: When the underlying older formations are represented by igneous or metamorphic rocks (unstratified rocks) and the overlying younger formations are sedimentary (stratified) rocks, the unconformity is called "nonconformity".

Angular Unconformity: When the younger and older sets of strata are not mutually parallel, then the unconformity is called "angular unconformity". In such a case, beds of one set (usually belonging to the older set) occur with a greater tilt or folding.

Disconformity: When the beds of the younger and older sets are mutually parallel and the contact plane of two sets is only an erosion surface, the unconformity is called "disconformity." Here, the lower set of beds would have undergone denudation prior to deposition of the overlying strata commenced.

Paraconformity: When the two sets of beds are parallel and the contact is a simple bedding plane, the unconformity is called "disconformity." In such cases, the unconformity is inferred by features like sudden change in fossil content or in lithological nature (such as the occurrence of coarse detrital sediments overlying fine sediments).

Regional and Local Unconformity: When an unconformity extends over a great area, it is called regional unconformity, but when it occurs over a relatively small or limited area it is called local unconformity.

Different unconformities are shown in Fig. 2.7 along with conformity structures for comparison. The MNO plane refers to the unconformity plane in these figures.

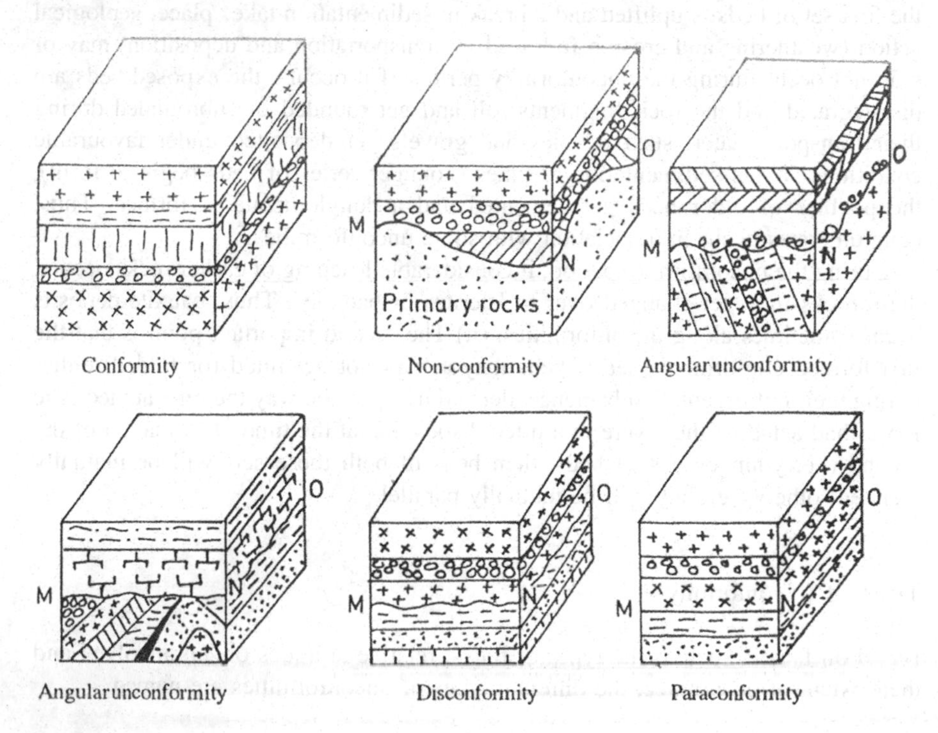

Fig. 2.7 Different types of unconformities

Recognition of Unconformity in the Field

Observations that help in locating an unconformity are as follows: (i) Difference in attitudes of two adjacent sets of beds (disconformity is an exception); (ii) Remarkable difference in nature, age and types of fossils in adjacent sets of beds; (iii) Occurrence of conglomerates along the unconformity plane; (iv) Occurrence of residual soil/laterite/bauxite along the unconformity surface; (v) Considerable difference in the degree of metamorphism of two adjacent sets of beds; and (vi) Stratigraphic correlation and lithological peculiarities.

The following enhancements help in locating an unconformity:

(a) ***Difference in the degree of induration***: the rocks on either side of an unconformity are likely to show different degrees of induration; the older rock showing greater induration.
(b) ***Differences in the grades of metamorphism***: the younger rocks are likely to be less metamorphosed than the older ones on either side of an unconformity.
(c) ***Differences in folding***: in some cases, the younger rocks will show less intense folding than the older on the two sides of an unconformity.
(d) ***Relation to intrusive***: in some cases, the presence or absence of an intrusive may determine the presence of an unconformity.

Unconformities are studied best in a single sharp exposure. Aerial photographs and satellite images are very useful in the study of unconformities.

Unconformity in Mineral Exploration—Localization of Ores

Since unconformity separates rocks of differing ages, it acts as barrier in mineral deposits associated with either of the two sets of rocks. Some difficulties posed by unconformities are similar to those faults, but the solutions seen in the case are different. Thus dying out of deposits at the plane of unconformity is common but, unlike in faults, their continuity also ends at the plane of unconformity. The plane of unconformity possesses of residually and mechanically concentrated mineral deposits such as bauxite, clay (in conglomerates) and gold. The plane of unconformities here acts as a channel for mineralizing solutions.

One problem is that the planes may be irregular and undulating, and the resulting mineral deposits are likely to be very irregular. This can be resolved by a systematic geological mapping followed by drilling in trace the continuity of the plane of unconformity itself. Proving of any deposit will come only at a later stage.

Most of the rock and geological structures described earlier offer controls in localizing mineral deposits. Some of the common structures which have a direct bearing on localization ore are (i) bedding planes, (ii) cleavages, (iii) joint planes and (iv) faults and folds.

Amongst these, the role played by fault and shear zones is important. The pre-ore faults in particular offer a ready passage to mineral and ore carrying fluids.

Faults and shear zones are widened when the ore or mineral fluids penetrate and move along the planes of weakness. Examples of shear zones acting as ore paths are very typically seen in deposits of barites (Pulivendala) in Andhra Pradesh, fluorite (Chandi Dongri), Copper (Khetri in Rajasthan) and lead–zinc (Zawar in Rajasthan). In all cases, the shear and fault zones have acted as field guides.

The localizing influence of fold is typically seen in the iron and manganese ore deposits. The synclinal troughs have usually shown a better concentration in many iron and iron and manganese deposits than anticlinal crests. The manganese deposits of Madhya Pradesh and Maharashtra show evidence of structural influences, particularly the influence of synclines. The influence of bedding planes, cleavage and schistosity is demonstrated by the mica deposits in Nellore and Hazaribagh where the mica-bearing pegmatites have intruded along the planes of schistosity and cleavage. This relationship has been used as a field guide.

Further Readings

Amaud Gerken JCD (1989) Foundation of exploration geophysics, Elsevier, Amsterdam, p 668

Billing MP (1972) Structural geology, EE edition. p 376

De Geoffroy JG, Wignall TK (1985) Designing Optimal Strategy for Mineral Exploration, Plenum Press, New York, p 364

Gass IG, Smith PJ, Wilson RCL (1979) Understanding the earth, ELBS, p 383

Ghosh SK (1993) Structural geology. Pergaman Press, p 376

Guilbert JM, Park CF (1985) The geology of ore deposits. WH Freeman, New York, p 985

Harris DP (1984) Mineral resources appraisal. Oxford University Press, Oxford, p 445

Kearey P, Vine FJ (1992) Global tectonics, Blackwell Scientific, Oxford, p 302

Klein C, Hurbut CS (1985) Manual of mineralogy (after J.D. Dana), 20th edn. Wiley, New York, p 596

Krauskopf KD (1967) Introduction to geochemistry. McGraw Hill, NY, p 721

Levinson (1974) Introduction to exploration geo-chemistry. Allied Publication (with 2nd ed. 1980 supplement), p 924

Marjoribanks R (2010) Geological methods in mineral exploration and mining. Springer, Berlin, p 333

Morse JG (2013) Nuclear methods and mineral exploration and production. Elsevier, Amsterdam, p 280

National Research Council of the USA (1981) Mineral resources: genetic understanding for practical application studies in geophysics, 119 p

Rankama K, Sahama TG (1950) Geochemistry. University of Chicago Press, Chicago, p 637

Rose AW, Hawkes HE, Webb JS (1979) Geo-chemistry in mineral exploration. Academic Press, New York, p 659

Rose A, Hawkes HE, Webb JS (1983) Geochemistry in mineral exploration. Academic Press, New York, p 657

Turner FJ, Verhoogen J (1960) Igneous and metamorphic petrology. Allied Pacific, p 694

Wellmer FW (1989) Economic evaluations in exploration. Springer, Berlin, p 163

Chapter 3
The Search for Ore Deposits and Chances of Success

3.1 Geological Models and Methods of Exploration

Every ore deposit has its special features that would the control choice of exploration norm. Moreover, no two deposits of the same mineral are identical. After the preliminary geological appraisal, if the mineral deposit appears to be promising, its exploration becomes an imminent step for a detailed evaluation of the deposit. Based on the specific need and other factors of cost, time and objectivity, an exploration strategy is to be chalked out. Consequent to the exploration, the processing of exploration data, and, assessment and evaluation is the next stages. The effective application of the geological knowledge to mineral exploration is the theme of this handbook. It depends largely on the availability of exploration methods: (a) locating ore deposits; (b) building up an adequate knowledge of location in which mining can be done and (c) assessing potential hazards such as minerals injurious to health.

The attempt is to carryout reconnaissance surveys which provide information on large tracks in most the economic/cost effective way and in the minimum time frame. This shall help to narrow the field investigation in the early stages of an exploration campaign. The detailed investigation of the identified selected sites is aimed to give definite information to recommend/commence for a detailed work. Reconnaissance methods adopted during this stage are given in Tables 3.1 and 3.2.

Geological, geophysical and geochemical methods to mineral exploration are important in locating ore methods, coal and hydrocarbon. These are specialized aspects and being useful information is available in several books such as Welte for hydrocarbon. Basically, Geophysical methods depend on measuring the physical properties of rocks. This has an advantage of giving information about rock, lithology and structure which are not accessible to direct observation. These methods include: (i) Seismic surveys; (ii) gravity surveys; (iii) magnetic surveys; (iv) electrical and electromagnetic method and (v) radiometric surveys. Remote sensing and imagery interpretation are also necessary, as given in Chap. 4.

G.S. Roonwal, *Mineral Exploration: Practical Application*,
Springer Geology, DOI 10.1007/978-981-10-5604-8_3

Table 3.1 Exploration procedures: an idealised sequence (for details of techniques see Chap. 8)

Decision to prospect, based on	Regional geology *favours* mineralisation of desired type *evidence* of mining in former times random finds of gossan or "shows" of sub-economic mineralisation
Regional reconnaissance by	Airborne geophysical surveys (magnetic, electromagnetic, radiometric, gravimetric) geochemical reconnaissance (stream-sediment or lake-sediment sampling) photo geological survey. Including use of satellite imagery ground geological reconnaissance, e.g. traverse mapping leading to: identification of favourable anomalies, staking of claims; random finds, staking of claims; or rejection of unfavourable areas, abandonment of project
Investigation of selected target areas	Geological mapping ground geophysical surveys (gravimeter. magnetometer, resistivity and induced potential surveys) detailed geochemical surveys (closely spaced stream-sediment sampling. sampling of drift, soil) exploratory pitting, trenching, trial boreholes leading to: discovery of deposits, staking of claim; identification of probable buried ore body; or abandonment of project
Assessment of ore body	Detailed topographic and geological survey of site further boreholes, logging of cares petrographical and chemical study of cores assaying of are samples leading to: decision to develop; decision to suspend operations; or decision to relinquish claim

Table 3.2 Assessment of potential ore-fields

Geological factors, size and structure of ore body	Principal metals, grade (i.e. concentration) of ore possible byproducts—(e.g. silver as trace metal enhances value of PbZn deposits)
Mining and metallurgical considerations	Mining procedure necessary. e.g. open-cast/underground
Extraction processes	Ease of separation of ore minerals, suitability for cheap treatment of purified are
Suitability of end product	Low levels of impurities such as Pin iron are desirable for use in steel making
Other considerations	Price of metal on world market, expected future demand ease of access, cost of development
Legislation relating to mining	Likelihood of future changes in social, political or economic climate liable to affect development

Geochemical and associated mineralogical investigations depend on both regional or reconnaissance and detailed surveys. The aim here is to locate an anomaly. Indicator elements or pathfinder elements are used for this. The geochemical methods need a direct collection of material. Soil samples, rock sample, soil gas and other gases, sediments of stream, lake, sea, and finally stream, lake or sea water. These are appropriately available in Ross et al. (1979), Levinson (1974, 1980). Deep sea mineral resources and exploration is important. Here, geological geophysical and geochemical methods are also used. This is a specialized aspect and one may refer to several books available, such as Cronan (1980), Rona et al.

(1983), Kunzondorf (1986), Roonwal (1986) amongst others. This was the time when oceanic minerals such as manganese nodules and massive sulphides had contracted being attention.

3.1.1 Efficiency of the Method in Exploration

The main aim of an exploration programme is to collect samples in desired quantum and types, and to prove the quality and extent of the deposit. Hence, it is important to adopt a method which yields such samples laterally and vertically. Samples need be made available continuously at regular intervals. In surfacial or shallow deposits, pilling and trenching yield most reliable sample. But, as mineralisation extends downward, most of the observations depend on diamond core drilling, which can later be substantiated by samples obtained by exploratory mining. This depends also on a host of conditions like organization of the team, nature of the country rock, stratigraphic sequence, nature of mineralization and structural features. Pitting in hard rock like quartzite is time-consuming and costly. It needs to be adopted to only under compelling circumstances. In such cases, core drilling will yield equally reliable data.

Cost of Exploration: Cost of operations need to be kept in mind. The money spent in exploration cannot be recovered in any form except in the sale value of the ore to be proved. This expenditure needs to be only to the level that can be easily absorbed later by the sale price.

Maintenance of Required Speed of Operations: Exploration programmes are mostly time-bound. Sometimes, either due to breakdown of machinery or any other reason, the programme may get delayed. Hence, the method chosen should be quite flexible with either spare machinery or manpower or by opting for substitution of the programme with a suitable alternative.

Topography of the Area: Some areas had remained inaccessible due to their altitude, thick vegetation, poor ground conditions like marshy terrain and lack of communication. Though man can reach in some of these areas, it would be very difficult to handle material and machinery. This particular situation is faced in case of some bauxite deposits. Even many iron ore deposits had been inaccessible in India until recent past. The mining geologist will have to adopt a method which entails transportation of man and material, availability of firm ground for setting up camps and drilling machines and availability of other infrastructural facilities.

Conformity with the Shape, Size and Pattern of Mineralisation: This factor almost controls the exploration operation. For deposits of large areal extent and surficial nature, pitting and trenching are recommended. Asbestos, bauxite, magnesite and diamond are explored by these methods. However, spacing of the pits or trenches is of vital consideration as it influences the total cost of operation. Care has, however, been taken in exploring some minerals like mica by pitting and drilling. For deep-seated deposits such as chromite, copper, lead and zinc ores, drilling will have to be carried out, drilling in such deposits is often controlled by structure and stratigraphic sequence.

Marketability of the Ore at the Exploratory Stage: In some exploration programme, it may be necessary to sell the ore raised during the operation. Here, the choice is in favour of a method which may yield large quantities of ore. This is generally the case with copper ore which is sent during exploration ore processing.

3.2 Exploration Procedure

Exploration activity is carried out in phases where activities are well defined from commodity and area selection to commercial production. Each phase aims to guide to prevent improper investment in the next stage. The different stages of activities and aims to be achieved during an exploration campaign are given in Table 3.1. Usually, an exploration activity is both costly and time-consuming. The cost-benefit risk ratio is very high involving serious financial constraints. To counter this problem, exploration programmes are taken up in stages, such as (1) preliminary exploration and (2) detailed exploration.

3.2.1 Preliminary Exploration

The preliminary exploration follows the regional mapping. At the stage of preliminary exploration of the mineral deposit, more detailed information is collected about the geological set-up and mineral occurrences. It may be carried out in a selected area with limited financial liability, considering the following factors:

(i) extent of the area to the explored
(ii) specific minerals/ores to be studied
(iii) type of data to be collected
(iv) time frame of exploration
(v) priority assigned to the work and
(vi) availability of fund.

Its objective is to generate necessary background information needed for decision on investment. Results of preliminary exploration may, however, just be sufficient to demonstrate if the deposit show promise to continuing investigation in the present technical and economic capability. At this stage, results are not encouraging at this stage, decision need to be taken if to abandon the project or delay it for future. It is necessary that such initial technical and economic assessment of an area is taken up with goal to prepare a pre-feasibility assessment report. Such a report needs to be brief, reflect the geological characteristics and clearly bring out the economic viability of the deposit, as given in Tables 3.2 and 3.3.

Table 3.3 Metalliferous mineral deposits: a geological grouping

1.	Deposits related to igneous processes orthomagmatic: segregated during consolidation pneumatolytic: associated with residual fluids exhalative and fumarolic: deposits of volcanic centres some hydrothermal deposits (see (4))
2.	Deposits related to sedimentary processes placers and related deposits: segregated by physical Processes metalliferous chemical sediments including deep-sea Deposits metalliferous residual sediments and weathering products some hydrothermal deposits (see (4))
3.	Deposits related to metamorphic processes pyrometasomatic deposits: formed at igneous contacts other metasomatic deposits of groups (1), (2) and (4) modified by metamorphism
4.	Hydrothermal deposits: formed through the agency of hot waters circulating in the host rocks: vein deposits replacement deposits

The results of preliminary exploration should help: (a) eliminate in inadequate/incomplete study; (b) avoid subsequent unwanted work if non-profitability is established in the initial evaluation of the economic soundness of the project and (c) determine the cost of the next stage of exploration. (d) Preliminary exploration throws up sufficient data to establish reliability of the information gathered on occurrence of the mineral deposit.

Other works that need to be carried out during preliminary exploration are as follows:

Preparation of contoured survey map, geological appraisal of the deposit, limited subsurface mapping, interpretation of borehole data obtained by drilling, chemical analyses of borehole and other samples and ore dressing tests. Geophysical and or geochemical prospecting may be useful in some cases.

Geological maps prepared during the appraisal shall provide additional information in the form of:

(a) geological sections,
(b) panel diagrams,
(c) isometric projections,
(d) isopach maps.

Drilling is an important activity at this stage. Based on regional geological mapping, required number of boreholes to be put and extent of depths to be decided on the basis of the degree of accuracy required. Drilling being is costly, enough care needs to be taken before it is executed. Following factors are to be carefully considered in case of drilling:

(a) type of ore deposit and rock formation,
(b) method of drilling,
(c) type of equipment to be used,
(d) core recovery and reliability of samples,

(e) analyses of core samples,
(f) electrical survey of boreholes and
(g) measurement of deviation and directional drilling.

Table 3.4 gives an overview of host rocks and associated ore mineral deposits.

Table 3.4 Element associations in mineral deposit, and examples of pathfinder elements

Rock type or occurrence	Elemental association	Pathfinder elements	Type of deposit
Plutontic association			
Ultramafic rock	Cr–Co–Ni–Cu	As	Au, Ag; vein-type
Mafic rocks	Ti–V–Sc	As	Au–Ag–Cu–Co–Zn;
Alakline rocks	Ti–Nb–Ta–Zr–RE–F–P	–	complex sulphide ores
Carbonatites	Re–Ti–Nb–Ta–P–F	–	–
Granitic rocks	Ba–Li–W–Mo–Sn–Zr–Hf–U–Th–Ti	Ba	W–Be–Zn–Mo–CuOPb;
Pegmatities	Li–Rb–Cs–Be–RE–Nb–Ta–U–Th–Zr–Hf–Sc	Be	Skarns Sn–W–Be; veins or greisens
Hydrothermal sulphide ores			
General Associations	Cu–Pb–Zn–Mo–Au–Ag–As–Hg–Sb–Se–Te–Co–Ni–U–V–Bi–Cd	Hg	Pb–Zn–Ag; complex sulphide deposits
Porphyry copper deposits	Cu–Mo–Re–Te–Au	Mo	W–Sn; contact
Complex sulphides	Hg–As–Sb–Se–Ag–Zn–Cd–Pb	–	Metamorphic deposits
Low temperature sulphides	Bi–Sb–As	Mn	Ba–Ag; vein deposits porphyry copper
Base metal deposits	Pb–Zn–Cd–Ba	–	U; sandstone-type
Precious metal	Au–Ag–Cu–Co–As	Se, V, Mo	U; sandstone-type
Precious metals	Au–Ag–TE–Hg	Cu, Bi, As, Co	–
Associated with mafic rocks	Ni–Cu–Pi–Co	Mo, Ni	–

(continued)

Table 3.4 (continued)

Rock type or occurrence	Elemental association	Pathfinder elements	Type of deposit
Contact metamorphic rocks			
–	–	Mo, Te, Au	Porphyry copper
Scheelite-cassiterite deposits	W–Sn–Mo	Pb, Cr, Cu	Platinum in ultramafic
Fluorite-helvite deposits	Re–F–R	Ni, Co	Rocks
Sedimentary associations			
Black shales	U–Cu–Pb–Zn–Cd–Ag–Au–B–Mo–Ni–As–Bi–Sb	Zn Ag–Pb–Zn	Sulphide deposits in general
Phosphorites	U–V–Mo–Ni–Ag–Pb–F–Re	Zn, Cu Cu–Pb–Zn;	Sulphide deposits in general
Laterites	Ni–Cr–V	Rn U;	All types of occurrences
Manganese oxides	Co–Ni–Mo–Zn–W–As–Ba–V	SO	Sulphide deposits of all types
Placers and sands	Au–Pt–Sn–Nb–TA–Zr–Hf–Th–RE–Ti		
Red beds, continental	U–V–Se–As–Mo–Pb–Cu		
Red beds, volcanic origin	Cu–Pb–Zn–Ag–V–Se		
Bauxites	Nb–Ti–Ga–Be		
Miscellaneous	K–Rb; Rb–Cs; Al–Ga; Si–Ge; Zr–Hf; Nb–Ta; RE; S–SE; Br–I; Zn–Cd; Rb–Ti; Pt–Pd–Ph–Ru–Os–Lr		

Note In most cases, several types of material (e.g. rock, soil, sediment, water and vegetation) can be sampled. In the case of sulphate only water is practical
Note: *RE* rare earth elements
Modified after Levinson (1972)

3.2.2 Detailed Exploration

The decision to continue to stage of detailed exploration requires an assessment of thepreliminary exploration results. Should the information collected generate positive inference, decision for detailed drilling is taken. It is expected to arrive at a degree of accuracy and confidence limit of the order of 90–95%. The expenditure on this generally varies between 3 and 5% of the total project cost.

The detailed exploration of mineral deposit is carried out for the following purposes: (a) to firm up the results of preliminary exploration, (b) to establish a higher degree of accuracy and dependability by projecting thegeometric pattern of the mineral deposit and blocking out reserves on a prior determined norms which would help in drawing up the mine plan, (c) to generate further data on mineral deposit with regard to mineability, overburden, management marketability, gangue minerals, associated minerals and mineral processing and (d) to help in preparation of a detailed project report for opening up a mine.

The detailed exploration, therefore, needs to include the following activities:

(a) contour mapping at closer intervals,
(b) drilling with borehole survey,
(c) aditing, trenching and blasting,
(d) collection of data on rock formations,
(e) preparation of large-scale geological map on I:2000 or I:1000 scale, cross-sections, etc., with reserve estimation,
(f) collection of borehole sample composite samples, and laboratory testings,
(g) collection of meteorological, ground water data and base line environmental data,
(h) mineral/ore processing and pilot plant tests.

3.3 Types of Deposits and Exploration

It is important to understand the nature of the mineral deposit, its mode of occurrence, the extent and complexity of its shape and size for deciding the type and quantum of exploration that may be necessary. GSI for example has classified Indian deposits into following types:

I. Strati-form strata-bound and tabular bodies with predictable regular habit.
II. Lenticular bodies and massive bodies of irregular shape and grade.
III. Lenses, veins and replacement bodies.
IV. Mineral deposits associated with reefs and veins.
V. Placer and residual refractory mineral deposits.

This is summarized in Table 3.5.

Various mineral deposits based on their shape, size, spatial distribution, structural disturbances may further be classified into (i) simple deposits, (ii) less complex deposits, (iii) complex deposits. The deposit falling in one of the types in the earlier classification may actually fall in more than one category depending on its complexity. The limestone deposit described under strati-form, strata-bound and tabular deposit may be a simple deposit, for example, Vindhyan limestone deposit of Katni-Satna area whereas limestone deposit which is structurally controlled may be either less complex deposit or complex deposits, like Walayar limestone deposit of Kerala and Dehra Dun limestone deposits of Uttarakhand. The norms of exploration would vary for different types of limestone deposits, depending on their complexity.

Table 3.5 Type of deposit and important mineral occurrences

I.	Stratiform strata-bound and tabular deposits	Coal seams, lignite bed, iron ore formations and cappings, manganese horizons in sedimentary and meta sedimentary sequence, thick bauxite cappings, chromite lodes in large ultramafics limestone, dolomite, barytes, gypsum, potash and salt bedschalk and fireclay, fuller's earth
II.	Lenticular bodies of all dimensions including bodies occurring in echelon pattern as silicified linear zones or composite veins	Base metal sulphides, supergene ironand manganese bodies in lateritic country, pockety bauxite and nickel Cobalt lateritoids, auriferous quartz reefs, graphite lenses: porphyry deposits of copper, molybdenum and tin, pyrite-pyrrhotite bodies
III.	Lenses, veins and pockets; stock work; irregular shaped, modest to small size bodies	Small multimetal complex sulphide bodies of Cu–Pb–Zn–Sb–Hg; podiformchromite; Sn–Ag chimneys and pipes; skarn deposits of scheelite, powellite, wollastonite, fluorite, etc., and semiprecious minerals, network of apatite, barytes, asbestos veins, vermiculite bodies, magnesite lenses and mica in pegmatite, pyrophyllite lenses and veins; opal, in situ sillimanite; kyanite lenses; high grade bauxite in clay pockets; clay, ochre and bentonite lenses, diamond pipes
IV.	Gemstone and rare metal pegmatite, reefs and veins	Tin–tungsten-tan talum-niobium molybdenum veins in pegmatite; beryl, topaz, emerald deposits, mineralisation associated with alkaline rock complexes and veins in carbonatite
V.	Placers and residual refractory mineral deposits of hill and valley wash	Placer tin and gold deposits, monazite, gamet, ilmenite, rutile, diamondiferous conglomerate; floats and gravel beds of corundum, kyanite, sillimanite floats and talus deposits of magnetite

(i) Simple deposit: Deposits are either strati-form, tabular with insignificant overburden or with a simple structure. The float deposits of various minerals may also be included in this category.
(ii) Less complex deposits: Deposits of regular shape and size but with structural manifestation and considerable overburden.
(iii) Complex deposits: Lode deposits, veins, lenticular deposits, structurally disturbed one with irregular shape and size.

Table 3.6 gives an overview of exploration programmes in such situations, whereas Table 3.7 gives exploration activity involved in different types deposits.

Table 3.6 Exploration methods: summary

Geophysical methods	Principal applications
Seismic surveys (reflection and refraction)	Elucidation of subsurface structure, especially: structure of sedimentary basins. Recognition of key horizons, unconformities. folds, faults in oil and gas fields: exploration of superficial deposits of construction sites
Seismicity records	Monitoring of active volcanic centres. Fault zones
Gravity	Elucidation of regional structure in sedimentary, igneous and metamorphic terrains. Useful at reconnaissance stage of exploration for minerals, hydrocarbons: identification of anomalies related to buried igneous centres, ore deposits
Magnetic	Elucidation of regional structure in igneous and metamorphic terrains. Useful for reconnaissance for mineral deposits; identification of anomalies related to buried igneous centres, iron formations
Electrical and electromagnetic	Resistivity surveys, principally to locate ore bodies and in borehole logging self potential (SP) and induced potential (IP) surveys to locate ore bodies electromagnetic surveys to locate ore bodies and detect rise of magma in volcanic centres
Radiometric	Prospecting for uranium, thorium; in areas of anomalous radioactivity
Remote sensing	Airborne geophysical reconnaissance: regional structure, topographical surveys, geological reconnaissance, surveillance of potential hazards such as volcanic centres, monitoring environmental changes
Subsurface sampling boreholes	Elucidation of regional succession and structure; mud logging for general lithology, microfauna logging of core for structural and petrographic detail (oilfields, mineral exploration, site investigation)
Auguring	Investigation of weathered mantle, superficial deposits
Geochemical and mineralogical methods reconnaissance surveys	Stream- or lake-sediment samples, water samples, reconnaissance for ore deposits, basis for investigation of geochemistry in relation to plant, animal or human health heavy mineral concentrates
Local surveys	Location of mineral deposits, investigation of possible geochemical hazards
Periodic sampling	Quality control (water and effluents): effects of pollution

3.4 Exploration Norms

Most of the mineral deposits for example in India are in small leasehold areas worked by private agencies. It is logical to expect that mining leases to come in future will also fall in this category. The small leaseholds may have about 20–50 ha area or even less with Pit's Mouth Value of very low value.

The factors which bring about variability in deposits include topographic setting, nature of mineralization, mode of origin, mode of occurrence, grade distribution within the deposit, run-of-mine grade, marketable grade and recoveries, etc. Despite the sevarieties, it should be possible to evolve certain exploration guidelines for a

Table 3.7 Exploration activity involved in these types of deposits

	Type of deposit	Exploration involved
(i) *Simple deposits*		
Group I	All float deposits, such as iron ore, manganese ore, bauxite, kyanite, sillimanite, etc.	Preparation of geological plans and sections. Pitting and trenching at suitable intervals, collection of samples and analysis of the same
Group II	All sedimentary deposits with minor/insignificant overburden, structurally simple ones; and placer deposits like limestone dolomite, gypsum, bentonite clays, limeshell, bauxite, silica sand, gold, beach sands, etc.	–
(ii) *Less complex deposits*		
Group I	Deposits with structural manifestations, such as iron ore, manganese ore, limestone, chromite, bauxite, phosphorite, calcite and asbestos	In addition to the above, drilling of boreholes depending upon the requirements and preparation of bore-hole logs based on sampling data
Group II	Deposits with considerable overburden like clays, limestone, kyanite, sillimanite, graphite, etc.	–
(iii) *Complex deposits*		
	Lode, deposits, veins, lenticular deposits and structurally disturbed one, such as base metals, mica, fluorspar, magnesite, gold, gemstone, etc.	In addition to the above, certain amount of exploratory mining with generation of detailed geological data particularly with respect to grade appraisal

Table 3.8 Norms of exploration for simple deposits

Simple deposit	Geological mapping	Pitting/trenching	Drilling	Sampling
(i) Group I	1:2000	Pitting 2–3 on each section line at 100 m interval	Not required	Pit sample
	1:1000	Trenching: if required, at 100 m	–	–
(ii) Group II	1:2000	Pitting 2–3 in each section line at 100 m interval trenching: if required	If required one or two boreholes	Chip channel, pit, borehole, bulk sample

unit area of the leasehold. The unit area that may be considered for the purpose may be 10 ha or its multiple. The exercise attempted is aimed at quantifying readily the exploration that may be carried out in the given area. The systematic geological mapping when carried out on a requisite scale would give fair idea of the type and quantum of exploration that may be necessary in the area. It is also expected that with the help of such norms at hand, it would be also possible to prepare the mining plan to be submitted along with the application for grant of mining lease. Norms for an exploration campaign of such cases is given in Tables 3.8, 3.9 and 3.10.

Table 3.9 Norm of exploration for less complex deposits

Less complex deposit	Geological mapping	Pitting/trenching	Drilling	Sampling/beneficiation test
(i) Group I	Surface mapping 1:1000	Pitting 4–5 in each ore zone or at every 50 m	1 or 2 boreholes in each cross-section as required	Chip, channel, pit/trench/borehole bulk sample
	1:500 Underground mapping (if required) 1:200 1:100	Interval, trenching: if required	cross-section at 100–200 m interval	Conducing of beneficiation test, if required
(ii) Group II	Surface mapping 1:1000 1:500 Underground mapping (if required 1:200 1:100	Pitting 4–5 in each ore zone or 100 m interval, trenching: not required	1 or 2 boreholes in each cross-section as required, cross-section at 100–200 m interval	Chip, channel, pit/trench/borehole, bulk sample, Conducing of beneficiation test, if required

Table 3.10 Norm of exploration for complex deposit

Geological mapping	Pitting/trenching	Drilling/exploratory mining	Sampling/beneficiation test
Surface mapping 1:2000 1:1000 1:500	Pitting and trenching, if required, 2–3 in each ore zone	2–3 bore holes in each cross-section at 50 more less interval. Exploratory mining up to 2 levels	Chip, channel pit, trench, borehold, sludge, bulk sampling, Conducting of beneficiation test on a pilot-plant scale

3.5 Analyzing Exploration Data

Regular assessment and review of exploration data are needed at all stages. The data generated requires an analysis to guide best results in minimum time and cost during exploration. The following steps are recommended for adoption in analyzing the exploration data:

1. Understanding the surface geology, regional and local structure, a broad idea of occurrence of mineral deposit, the geological set-up of the project area and its relation with the regional structure for delineation of the ore zone.
2. Assessing the ore intersections obtained during surface trenching, potting and drilling to understand the extent of deposit.

3. Geological data from the underground exploration by drives, cross-cuts, shafts and winzes need to be utilized. The underground geological mapping shall define the shape and size of the ore zones. It shall reveal details of structural aspects. This needs in delineation of small ore shoots.
4. Such data shall help in correlation between ore zones intersected by surface drills and in underground exploration.
5. Estimate the most suitable cut-off from geological and economic consideration. The check exercises on natural cut-off on different sets of samples are done by plotting the frequency distribution of assay values from the full width. One may select two or three economic cut-offs around geological cut-off so that it can provide data about alternative widths and grade for further assessment. An evolution on alternative economic cut-off helps at a later date, depending on changes in price, in making suitable adjustment in stope boundary. When a project turns out to be unviable, the grade needs to be improved, at the cost of tonnage to convert it into a viable one. Trials on two or three cut-off, at initial stage, help in conducting the feasibility.
6. A clear assessment of the shape of ore zones and their distribution pattern, guides in projecting probable pinching thickening or thinning, beyond already explored extension of the drill hole intersection, cut-off limits are worked out keeping in view the structural trend, the projected level plans of the are shoots at various levels.
7. Appraisal report in absence of exploratory mining data, gives a uniform influence to the intersections of the drill hole available on a grid pattern. It gives an impression that the ore body is tabular. Subsequently available exploratory mining data, if the ore body turns' out to be of lensoid or elliptical in plan or section, an overall tonnage is worked out. Sometimes for rectangular and ellipsoid shapes, there can be a difference in volume, as much as 30–40%. Therefore, a careful review of available exploratory data and check up examination guides to satisfy the shape and size of the ore body. The estimation parameters earlier assumed by the exploring agency and the result of exploration are complementary to each other.
8. The exploration campaigns usually carry out requisite exercises for determining tonnage factor. However, in respect of multi-metal deposits, it may be required to compute tonnage factors separately for different grades.
9. Based on a given set of data, different methods exist to estimate reserves. The procedure differs from deposit to deposit as well as for the type and intensity of exploration done. This estimation of reserves depends upon the degree of reliability on the exploratory and geological information. In case of detailed

mining plan, levelwise and oreshootwise reserves are studied. The grade is based upon the lateral extent of mineralisation in a particular level and vertical extent of influence, at halfway to the level above and below. When, the ore body is regular. Cross-sectional method of estimation, levelwise and lensewise is adopted.

10. An exploration evaluation depends on adequate sampling points. The method of sampling adopted, the nature of the mineral deposit.
11. The geological data when processed brings out significant gaps in exploration. It is likely that where geologically ore zone continues, there may be a doubt whether the ore body continues in strike and dip. In other cases, there is a possibility of addition to the reserves by proving extensions.

In different data-processing stages knowledge of ore mineralogy, host rock, structural features including folds, faults, shears; joints and controls of mineralisation within the given set up is needed. General guidelines for exploration of some important minerals are given in Tables 3.11, 3.12, 3.13, 3.14, 3.15, 3.16, 3.17, 3.18, 3.19 and 3.20.

Table 3.11 Guidelines for exploration for iron ore

	Method deposit	Capping type deposit	Reef type	Remarks
1.	Mapping Underground mapping for adits	1:1000–1:2000 1:200	1:1000–1:2000 1:100–1:200	To map lithology and boundaries of soil, ore and waste, structural features, etc.
2.	Drilling	50–150 m section interval, 2–4 boreholes in each bottom of the ore, section to outline	100–150 m section interval down to 90 m depth	Sludge collection is important wherever core loss occurs. Dry drilling useful in Soft zones
3.	Pilling	Deep pits down to 15 m depth, 2–4 nos. on each section line to determine lump to fine ratio, etc.	1–2 in every third section	–
4.	Aditing	Cross-cutting adit intersecting ore body, 2–3 nos. along representative length	2–6 adits at different levels and at 300–500 m lateral interval, sections with 30–50 m	–
5.	Sampling	Cores and sludge from drills, for every 1–2 m interval. Bulk from pits and adits	Core and sludge from drills. Bulk samples for every 2 m depth from pits and adits for grade and size classification	–

Table 3.12 Guidelines for exploration for manganese ore

	Method	Simpler deposits type-IM.P. & Mah.	Complicated deposits type-2M. P. & Mah.	Lateritoid deposits type-3
1.	Geological mapping	1:1000 1:2000	1:1000 1:2000	Due to high order of variability, the
2.	U.G. mapping	1:200	1:100	Chief mode of exploration is potting and trenching and at
3.	Drilling	100–200 m section interval at 3–4 levels	25–50 m section interval at 3–4 levels	Places shallow drilling. Review of exploratory work at every stage is desired due to usually low potentiality
4.	Sampling	Core and sludge, channel and blast	Core and sludge, channel and blast	–

Table 3.13 Guidelines for exploration for chromite

Method	Stratiform deposits without complex structure	Podiform bodies with irregular pinch and swell or grade	Remarks
Mapping	1:1000	1:1000	To isolate ultra-mafic and chromite bodies, shear zones and zones of serpentinisation
Drilling	100 m interval	30–100 m section interval	Also to study core recovery in the ore zones
Drilling	100 m interval	30–100 m at ¾ zones, levels workable depth	Lithological set up and occurrence of unsuspected ore bodies, etc.
Trenching/pitting	3–5 for every ore body lens	3–5 for each ore body	More helpful in vein and lens-like bodies. Recovery with depth should also be studied
Sampling	Core, sludge	Core, sludge and chip/channel samples from pit	–

Table 3.14 Guidelines for exploration for mica

1.	Mapping (Surface) (Underground)	1:1000 1:500 1:100–1:50
2.	Trenching	At 30 m strike interval
3.	Exploratory mining	8 m interval levels, winzes and raises and cross-cuts at close intervals
4.	Drilling	Extension rods fitted to jackhammer and by other methods in between cross-cuts and advancing faces

Table 3.15 Guidelines for exploration for gold

1.	Mapping surface Underground mapping	1:1000–1:2000	Most gold deposits are however explored by the same methods applicable to the sulphide deposits
2.	Surface drilling	50–100 m grid, drilling in 2–3 levels, 30–60–90 m vertical interval to trace and intersect mineralised zones	–
3.	Underground drilling	As and where necessary	–
4.	Exploratory mining	3 ore more levels over the entire/part strike length of ore body at 30 m level interval and winzes along suitable intervals	–
5.	Sampling	Core and sludge, blast and channel, bulk sample from underground developments for beneficiation test	–

3.6 Economics of Exploration and Mining

In all cases of mineral exploration be exploratory drilling, and subsequent mining, cost need to be kept in mind. Exploration is a capital investment. Hence the cost of exploration has been added, totally, partially, either as lump sum payment or on differed payment basics, to capital cost. High interest charges on capital, during the construction period of mining project can act as a millstone round its neck! A visible project can become uneconomical. On the other hand, it should be kept in mind that stinting on exploration can prove to be disastrous, in a long way. This could lead to delay in commencing the project or even abandoning the project after incurring considerable expenditure. Therefore, though it may look out of place, money spent judiciously on exploration is money well spent.

An expenditure amounting to 2% of the total investment would seem justified for a good project.

We all agree that together with basic knowledge in an exploration programme to locate a mineral deposit, is to mining it out. This needs to be done on the premise that it should lead to resource for the society as also generate wealth to company and to the country.

Therefore, an exploration geologist is not only looking at structure, ore mineralogy and petrography of the host rock, but necessarily a techno—economic feasibility of the deposit which can be developed as a viable project for mining. It is important to accept that no venture or company can service for long term basis if it is not profitable, and mining and exploration for mineral deposit are very much part of this. Some basic knowledge of economic and commercial aspect is thus necessary to know. An exploration geologist need not be a combination of all discipline of economics and commerce. He is naturally expected to understand the

Table 3.16 Guidelines for exploration for copper, lead, zinc and multi-metal combinations

S. No.	Method	Thin low dipping strata-bound body with a little structural control	Large lenticular replacement body	Low to moderately dipping large bodies with simple structure	Steep dipping small ore bodies structurally complicated	Remarks
1.	Mapping	1:10,000–1:2000	1:1000	1:1000–1:2000	1:100	To estimate
	Underground mapping	1:100–1:200	1:100–1:200	1:100–1:200	1:100–1:200	Footwall hanging along strike, individual lenses and vein, assay boundaries, etc., lithological units, old working, abandoned mines dumps, etc. For establishing depth continuity strike extension, width of mineralisation and depth of oxidation
2.	Drilling	50–120 m grid drilling with underground drilling, if necessary	50–120 m section interval intersection at ¾ levels at 50–60 m vertical interval with underground drilling, if necessary	100–120 m section interval. Intersection at 3–4 levels with 50–60 m intervals, drilling, if necessary	50 m section interval intersection at 4–5 levels at vertical intervals at 30 m	
3.	Exploratory mining	1 km strike length of the deposit by riving cross-cutting, etc.	The entire strike length in two levels by drives, winzes and arises	Entire strike length in two levels by drive with cross-cut at 30 m interval and winzes and raises	Entire strike length in two levels by drive with cross-cuts at 15 m intervals with raises and cross-cuts	
4.	Sampling	Core and sludge underground channel samples bulk samples for beneficiation test	Core and sludge, channel sample, bulk samples, etc.	Core and sludge, channel samples and bulk samples	Core and sludge underground channel	–

Table 3.17 Guidelines for exploration work for lead, zinc and copper

	Method	Simple-type deposit	Complicated-type deposit
1	Geological mapping	1:1000	1:1000
2.	Drilling	100–200 m grid	100–150 m section interval, intersection at at 2–4 levels
3.	Pitting/trenching	4–6 numbers per km^2/ trenching not recommended	Pitting is not recommended, trenching at 100–150 m section lines
4.	Exploratory mining	Not recommended	Two-level development by drives/winzes, cross-cuts at approximately 30 m interval
5.	Sampling: core and sludge, blast and channel samples and bulk samples for beneficiation		

Table 3.18 Guidelines for exploration for limestone and dolomite

Type of limestone deposit				
	Method	Simple	Complicated	Highly complicated
1.	Geological mapping	1:1000 1:2000	1:1000	1:1000
2.	Drilling	200–300 m grid	100–150 m grid	50–100 m grid
3.	Pitting/trenching	4–5 per 10 ha	5–8 per 10 ha	10–15 per 10 ha
4.	Sampling	Core, blast, channel, chip	Core, blast, channel, chip	Core, blast, channel, chip

Table 3.19 Guidelines for exploration work for bauxite

Method	Type-I Bedded extensive	Type-2 Lenticular extensive	Type-3 Erratic patchy
Geological mapping	1: 1000	1:1000	1:1000
Drilling	100 m grid/interval	50 m grid/interval	Not recommended
Pitting/trenching	100 m grid/interval	60 m grid/interval	25 m grid/interval
Sampling	Core, sludge sample channel sample 20 m interval along scarp, channel, (all 4 walls recommended) for pit, bulk sample	Core, sludge channel at 20 m interval along scarp channel (all 4 walls recommended) for pit, bulk sample	Channel sampling for pits (all 4 walls recommended) bulk sample

interdisciplinary nature of this programme. It is also true that one can generally handle small scale of mining projects, but for a large scale venture there is always a need for specialist on economics and commerce who shall guide the financial planning of the project. No project can commence without the proper feasibility and any decision on investment will depend on the basic composition of the deposit and its resource potential on terms of economic returns.

Table 3.20 Some universal and local plant indicators with their associated metals

1.	Cobalt	*Crotalaria cobalticola, Silene cobalticola*	–
2.	Copper	*Acrocephalus robertii, Astragalus declinatus, Becium Homblei, Gypsophila patrini, Merceya latifolia, Merceya ligulata, Mielichhoferia macro-carpa, Mielichhoferia mielichhoferi, Tephrosia* sp., *Viscaria alpina*	*Armeria maritima* *Elsholtzia haichowensis* *Eschscholtzia mexicana* *Polycarpaea glabra* *Polycarpaea spriostylis*
3.	Iron	–	*Betula* sp. *Clusia rosea, Dacrydium caledonicum,* Damnaraovata *Eutessa intermedia*
4.	Lead	–	*Baptisia bracteata* *Erianthus giganteus* *Tephrosia polyzyga*
5.	Manganese	–	*Digitalis purpurea, Fucus vesiculolus, Trapa natans, Zostera nana*
6.	Molybdenum	–	*Astragalus declinatus*
7.	Nickel	–	*Alyssum bertolonii, Asplenium adulterium, Pulsatilla patens*
8.	Phosphorus	–	*Convolvulus althaeoides*
9.	Selenium	*Aster venusta, Astragalus* spp., *Oonopsis* spp., *Stanleya* spp.	*Neptunia amplexicanlis*
10.	Selenium and uranium	Astragalus (Certain spp.)	–
11.	Silver	–	*Eriogonum ovalifolium, Lonicera confuse*
12	Vanadium	*Astragalus bisulcatus*	–
13.	Zinc	*Thlaspi calaminare, Thlaspi cepaeacfolium, Viola calaminaria, Viola lutea*	*Gomphrena canescens, Matricaria americana, Philadelphus* sp.

For this, it is good to separately consider (1) exploration—mining and (2) mineral processing/beneficiation. Indeed there are after basic needs for success—infrastructure is one of the important parameter. Without support of rail-road and related aspects mined out mined cannot be transported. In keeping with the details of deposit and technology needed to mine it, costing needs to be understood on

(a) Geology and exploration; (b) Mine development.

How to Calculate/estimate mining cost—example

In producing mine for copper ore, per annum base:

Cost of the ore—1.5% Copper
The mill grade—2% Copper,
Annual production—200,000 tonnes/per year

(A) **Mining Costs**
I. **Mining costs**: This would include.

- Labour, supervisor staff and administrative staff
- Costs of exploration and blasting
- Costs of support such as timbering, roof bolting, pack wall
- Drainage and pumping
- Ventilation and related aspects
- Haulage and transport
- Maintenance
- Stores and supplies

This is in addition to mine development—sinking of shaft

II. **Indirect costs**

Royalty @ *x*/tonnes of ore.
Office and other expenditure not covered above.

(b) **Total Costs**: mining and transportation of ore, Beneficiation and ore—processing, Smelting and refining
(c) **Revenue**: Assuming a mill efficiency of 90%, and smelter efficiency of 90%, then copper metal recovered to shall be

$$200{,}000 \times \frac{2}{100} \times \frac{90}{100} = 3600\,\text{per/annum}$$

Thus taking a price of control say at Rs. 12,000 per tonnes, the revenue per annum is Rs. 43,200,000 (this could change with fluctuation in metal price).

(d) **Deductions and Contribution**: Depreciation or an amortization, Interest on loan capital @ say 8%, Interest on cooking copper @ say 8%, Net realization/revenue or per profit = (C)–(B)
(e) **Profit**: The net project may be shown as a percentage of the equity capital (*E*).

$$(E) = \frac{(C) - (B)}{(E)} \times 100$$

It is important to work out the profit of the project using the known techniques.

(a) Payback method
(b) Accounting method
(c) Discounting costs float method
(d) Present values index method

Amongst the method listed, (c), (d), (e) method is generally adopted in evaluating the profitability of a deposit and its mining.

It is normed in a mineral—mining project that the project does not earn any revenue profit say for 3–5 years. During this period there is only costs "outflow" expenditure. After the mining operation begins, the Project starts to "earn" which means cash "inflow". Assuming a certain number of years to mine out the mineral (life of mine project), the cash outflow is counted are revenue/expenditure planned for the future.

Capital Cost

In most case capital (money) has to be raised through share capital (equity), and loan capital, a fair approximation of the total requirements is necessary. It is to be noted that dividend (profit) to the share holders has to be declared at a reasonable rate present, on equity share, interest paid on the loan, etc. Besides, items like machinery, equipment, building—their depreciation/amortization need to be accepted.

The following is the check list of items which are included in capital cost:

- Cost of land or rent
- Investment on infrastructure such as road, water supply, etc.
- Cost of over burden removal
- Development activity—shaft sinking, etc.
- Sampling and ore—analyses
- Ventilation, drainage, etc.
- Power supply
- Workshops
- Transport
- Building office, store, etc.
- Residence for employees
- Social aspects, hospital, club, child centre care centre, school, centre

These are only broad guidelines. Minor adjustments are required according to mineral, nature of deposit and market consideration.

3.7 Organization and Methods in Mineral Exploration

Mineral exploration aims at searching out new mineral deposits. In a virgin terrain, this may involve the location of possible targets by prospecting. In an already known area or in a developed mine, exploration may be done solely to study the potential of a new mining block. The object of such efforts is to locate and develop more mining blocks in the minimum of time and cost. Exploration is in short an important economic function of the mineral industry, its main aim being the creation of new profit centres for tomorrow. In the national context, however, in some cases, it may not be a centre of profit monetarily, but a national need.

Organization

Prospecting and exploration have for long been dominated by individual efforts. This has been true throughout the world. In India also, in a sense it has been true. Exploration has become a business activity of the mineral industry only in recent times and the concept of management is in a state of development even in very advanced countries. In India, the management concept is yet to develop fully.

Proper organization and management are, however, essential ingredients in the successful execution of exploration. The major function of exploration management is to coordinate the three major factors required for locating new deposits, viz., ideas, money and expertise. Of these, the factor of luck is the major imponderable in every exploration effort. By quantifying and eliminating various unknowns, the factor of luck can be reduced to some specific risk level which can be foreseen and measured.

Prospecting for new deposits, apart from exploring known ones, is a major gamble. The law of "gambler's ruin" applies here too. The rule expresses the charges of going broke in a short run of bad luck. Such spells of bad luck can be compared to the non-discovery of any new deposit in a continuous series of search in an apparently favourable terrain. The law of "gamble's ruin" suggests that, in order to overcome such a row of failures, it is essential to keep trying despite failure. Such decisions are possible only when large capital is available and the decision to continue is based on a logical and scientific reasoning. Scientific reasoning in this case comes only from geological knowledge.

It is logical therefore that exploration management has to be in the hands of a geologist. An exploration manager should combine broad geological knowledge with imagination, physical endurance, tenacity of purpose, readiness to assume risks and should be prepared to take decisions quickly in the best possible way, in many cases even without knowing all the facts. It is also important that exploration management has to be as close to the field as is feasible, the authority for making technical decisions in particular resting at the field level. The management functions of mineral exploration can be broadly identified in the following types of activities:

(1) Selection of minerals for exploration
(2) Acquisition of mineral rights
(3) Recruitment and organization of personnel
(4) Procurement of equipment and
(5) Co-ordination and administration

Selection of Minerals for Exploration

The process of selection is essentially guided by market conditions for a particular mineral or mineral-based industry. Thus, a cement plant requires limestone, a steel plant needs iron ore, limestone and dolomite and a ceramic industry look for clay deposits. The demand may be for the export of raw materials like iron and manganese ores. The organization or geologists entrusted with the task of prospecting and exploring for any deposit need to know the type of ore, quantities, specifications, location and also the rate at which the ore materials are needed. Normally, in such cases, the choice of the mineral/ore is outside the control of exploration management. The specific needs are conveyed to the exploring agency by the industry which is looking for the specific ore or ores.

The exploration geologist/organization may have a purely commercial aim in finding and developing ore bodies to attract interested investors to develop them commercially. In such cases, it is essential to study the market conditions, and select minerals which are easy to locate and have a ready market. Here, the selection of the mineral is largely in the hands of the exploration organization. When the ore indication is shown, one needs to gather information where to look it. One could go through memoirs, records, bulletins of the Geological Survey of respective countries. Commonly the geological survey agencies, and other companies that deal with exploration and mining of mineral industries, are the sources of geological information in other countries.

Acquisition of Mineral Rights

In India, anyone who wishes to undertake mineral exploration would be legally required to possess a certificate of approval from the State Government and then obtain a prospecting licence or mining lease. Mineral discoveries are made during routine geological work like systematic geological mapping or other exploration work done by the Geological Survey of India or the State Departments of Geological and Mining. Such discoveries can be studied from a scientific angle without disturbing the surface, but for chipping pieces of rock, without recourse to any level sanction. However, for mineral exploration for purposes of opening up and mining mineral deposits, certain legal sanctions are required. The procedures to be followed are embodies in the various statutes of the Central and State Governments. For acquiring the mineral rights of an area, the pre-requisites in stages are (i) Certificates of Approval, (ii) a Prospecting Licence (PL) and/or a Mining Lease (ML) if already issued, (iii) an incometax clearance certificate from the Incometax officer concerned and (iv) a valid clearance certificate of payment of mining dues such as royalty, surface rent, etc. In many countries, a clearance certificate from the state environment agency is required.

(i) Certificate of Approval: Before venturing to prospect or exploit any major mineral, an entrepreneur must possess a Certificate of Approval. This certificate is issued by the concerned State Government and signifies the financial and/or technical ability to commence mining.

(ii) Prospecting Licence: A prospecting geologist should be well conversant with the procedure for acquiring the rights to prospect and mine mineral deposits. It is essential to obtain a P.L. for the areas to be investigated. Generally, a large area should be chosen as a prospecting target which can be progressively reduced after locating some promising mineral deposits. Finally, the mining lease may be taken only for such areas which can sustain mining.

A Prospecting Licence is issued for a period of 1–2 years, which is renewable for the additional year. During this period, the licence holder is expected to prove the deposit in order to enable the opening up of a mine.

(iii) Laws governing mining lease: A Mining Lease may be taken directly in areas where the presence of mineral deposit is known. Or it may be taken after prospecting of a few targets in an area of promise. A mining lease area should cover the deposit and some adjacent areas for the development of ancillary facilities like waste disposal, construction of surface structures such as office, colony, explosive magazine, processing plants and space required for other facilities connected with mining. The mining lease is issued for periods of 20 or 30 years and can be renewed for a similar period. The lease provides exclusive rights to the lessee to exploit, process and market the ore and ore products.

3.8 Recruitment and Deployment of Personnel

The success of any exploration venture depends on the training and background of its personnel. The recruitment of exploration personnel should be done carefully giving due emphasis to the academic background and qualifications, professional experience and scientific temperament. In addition, an exploration geologist should have the following qualities:

(1) Ability to keep track of various technical developments in the subject.
(2) Ability to keep in touch with organizations or persons who can provide the best possible information on a variety of geological and allied subjects like: (i) geochemistry, (ii) geophysics, (iii) drilling and exploratory mining, (iv) mining methods, (v) ore dressing, (vi) metallurgy, (vii) mineral industry and trade, (viii) economics and financing and (ix) mineral and taxation laws, etc.
(3) Ability to impart proficiently various necessary skills to the less trained persons. This includes the training of people in specific practical skills like the excavating of a pit or trench, cutting of channels for sampling, sampling and sample preparation, etc.
(4) Ability to report lucidly, legibly and in time.
(5) Ability to negotiate with various State and Central Government agencies and other parties on technical and, non-technical details like acquiring land, establishing camps, arranging of provisions, services, security, etc.
(6) Ability to establish good contacts with the local community.
(7) Ability to foster a spirit of co-operation among the members of the team.
(8) An open mind free of prejudices.

Field Organization

Since all, or most, exploratory efforts do not assure any immediate financial, returns, the field organization should be small and should be efficiently managed. The field party should be assisted by a few trained and skilled persons who can help in survey, sampling and other work. Apart from this, there will have to be some office help to keep track of the day-to-day running of the organization, maintain accounts, records, etc. There may have to be a few guards to take care of the equipment, records and camps. Other personnel may include cooks, water carriers and labourers to do heavy manual work. Coordinating such units for the general success of the exploration effort is a skilled managerial job. Each unit should be assigned the job for which it is best suited and should work according to the general time schedule.

The administration of any exploration unit should be organized to keep track of all activities, arrange for the supply of equipment and materials and relieve technical personnel from routine details. Speed, efficiency and proper coordination and accountability should be the keynote of administration. Some of these items of work are beyond the general sphere of activities of the individual geologists but they are being mentioned here to keep them well informed about organizational details.

An exploration party may consist of many people, depending upon the extent and intensity of the work involved. The choice is left to the individual geologist who should be able to choose and deploy personnel according to the specific job requirement. In addition, the services of specialists like photo-geologists, geochemists, geophysics, mineralogists, ore dressing engineers, etc., may be deployed as consultants when so required.

Camping: Camping for short duration is not a major problem in most parts of India where some kind of public accommodation in the form of rest houses, tourist homes, etc., are available. Even in villages, temporary accommodation can be fairly easily arranged.

In cases where the prospective area is deep within virgin terrain, it would become necessary to establish a temporary camp. The camp site should be chosen as near to the target area as possible and should have a source of clean drinking water nearby. It would be preferable to camp as near a village as the circumstances permit so that provisions, labour and communications are easily arranged. The erection and maintenance of some temporary structures to house the personnel and equipment would be necessary in case of such camps. In the conditions obtaining in most parts of India, it would be best to erect mud huts with straw/grass or galvanized iron sheet roofs. Tents are useful if their occupation is confined to the dry months only.

A camp in remote areas should have a first aid kit and also some patent medicines. The geologist who heads the operation or someone as responsible should have training in first aid and also in the administration of a few drugs. This will normally ensure the availability of first aid and a necessary minimum of medical attention, for the personnel. It would be advisable to arrange a source which can periodically replace provisions.

3.9 Procurement of Equipment Needed in Mineral Exploration

A large variety of equipment is needed for mineral exploration purposes. Of these, some are for constant and continuous use and should be purchased for permanent retention. Others may be needed only occasionally and might be hired when a specific need arises. Some of the items of equipment which are of constant and continuous use are dealt with below:

1. ***Base maps***

A well-prepared topographical and geological map of the area which is to be investigated is the most important ingredient of a prospecting expedition. The plan should show topographical and geological details on sufficiently large scale. In India, the standard scales for systematic geological mapping are (presently 1:50,000).

Where a geological map is not available, a geologist may have to map the area first. For this, topographic base maps are required. Such maps can also be obtained from the offices of the Survey of India. When topographic maps are not available, use may be made of forest maps, revenue and cadastral maps available with the district revenue and/or forest authorities. When no base maps are available, it has to be prepared by the Surveyor first if a surveyor is available, or else, a geologist himself should be able to prepare one.

Where necessary, maps have got to be enlarged systematically on to a suitable scale before using for geological mapping. Square method, protractor and proportionate compass method, photographic methods are commonly used depending on the urgency and availability of the facility. It is also possible to scan the maps and enlarge them digitally, using a common graphical software. In the case of areas already worked, it is possible that some plans and other details may be available with government agencies like Indian Bureau of Mines. Where legally feasible and technically relevant, the possibility of these being used deserves to be considered.

In the field, a map comes to be used rather roughly. Precautions are necessary to prevent the map from getting torn and damaged. The best precaution is to mount the paper map on a cloth backing and bind it with card board covers in such a way that they can be loosely folded into a book. Facilities for map mounting and even mounted maps are available at the map sales offices of the Survey of India. Such a book covered with a thin transparent plastic sheet would completely protect the map from most of the damages normally encountered in the field. A permanent map case of suitable shape and made of light, rust-free metal would also be needed for safe transit and storage of maps and plans.

Satellite images and aerial photographs

Satellite images and aerial photographs are useful for structural, lithological and alteration mapping. Different lithologies are depicted in different colours in coloured satellite images. A geologist can save these images in his/her Tablet or

GPS and use it in a scale desired. If aerial photographs are used, a pocket stereoscope is a great help to see the geological features in three dimensions.

2. ***Compass***

A compass is used for finding directions, taking traverses and locating one's own position in the map. The magnetic needle of the compass always points towards the magnetic (and not the true) north. A compass meant for a geologist's field work usually has some built-in arrangement for measuring strikes, dips of bedding planes and inclinations of the various planar features. For versatility in use, a Brunton Compass is the best. A Brunton compass is designed in such a way that it can be used as compass, clinometers and hand-level. For most uses, this instrument can be held at one's hand. When accurate measurements are required, it may be mounted on a tripod stand. Brunton compass may be used for making the following observations: bearings, vertical angles, strike and dip, geological mapping, elevation calculation, etc.

3. ***Global Positioning System (GPS)***

It is often necessary to know the accurate coordinate position of an outcrop, sample location, bedding plane measurements, etc. A handheld GPS can provide the geographic location with an accuracy of ±few metres. In addition, a geologist can upload the necessary maps, satellite images and aerial photographs and use them while navigating in the field. This device is also useful for finding any point desired and finding the right path back to the camp.

4. ***Hamer***

Hamer is essential for breaking rocks and collecting samples. Several types of geological hammers are available; the most useful of them may have a square hammer on one side and a chisel edge on the other. A hammer of this type with a longer handle than usual may be used as a pick in climbing steep inclines. A light wooden handle with some flexibility or a steel or wooden handle with shock absorbing material will protect the striker from shock. The usual field hammer may weigh 0.5–1 kg.

5. ***Chisel***

Chisel is necessary to wedge out rock samples as also in cutting channels for sampling for collecting chip samples. A chisel-like tool with a pointed end is termed a moil and is useful in cutting and chiseling out rock specimens.

6. ***Magnifying glass***

A pocket lens with a magnification of 10 times is useful for most of the field examinations. Special large diameter lenses with a large field and high magnifying power may be required when working in a multimineral area. For purposes of reading maps, a map reading hand lens will be useful.

7. ***Magnet***

A horseshoe or bar magnet of small size is essential for testing magnetic minerals.

8. ***Measuring tape***

A geological field party should have at least one steel or metallic (cloth) tape of 30 m. length and another pocket steel tape of 2 m length. The first for ground measurement, measuring traverse lines, etc., and the other for measuring in shorter units, such as thickness of veins, beds, etc. The survey teams require 60 m steel tape as well as pocket steel tape of 2 m length.

9. ***Protractor***

A rectangular protractor scale made of plastic and of convenient size to be carried in one's pocket is most advisable. For purposes of plotting and measuring the bearings, the protractor may be circular showing 0°–360° divisions and made of good transparent plastic.

10. ***Field notebook***

A field notebook may be about 18 × 12 cm in size containing some 200 pages. Most of the pages should be unruled. Some graph pages and a few ruled pages should be included at intervals. Some detachable perforated plain sheets at the end of the book will be of immense use. The book should have a hard cover of paste board or plastic material and should show the name and address of the geologist, details of the project, and date of operation. A field notebook may be provided with a unit conversion table, equal-area stereographic net, natural trigonometric functions and a nomogram for bed thickness calculation.

In addition streak plate, glazed porcelain pieces and pocket knife should be available with the filed party.

Survey work requires special field books to put down all the recordings made in the field.

11. ***Pickaxes, shovels, baskets, etc.***

When some minor excavations become necessary to study an outcrop, pickaxes, shovels, etc., may be needed. These may be chosen from the wide variety available in the market.

Other essential items are mostly non-specialized equipment and may be chosen from among the various standard brands available.

12. ***Transport***

Some form of transport is essential in exploration camps. Petrol/diesel driven jeeps and trucks are the most versatile forms of transport. However, if such transport is too costly or cannot be used, and for small-scale operations, animals or animal drawn vehicles like bullock carts could be used.

3.10 Methods in Exploration

Mineral exploration consists of two activities. One is prospecting which, in essence, is the choosing of suitable mineral targets for exploration. The other is exploration which is the proving of targets located during prospecting. The methods employed during either of the activities cannot be rigidly isolated. However, certain types of activities are typical of both and can be broadly recognized. But prior to starting prospecting or exploration it is necessary to choose the target areas which can yield mineral deposits. The methods of choosing such target areas are discussed below.

Methods of Choosing Target Areas
The selection of target areas should be done by a process of sequential elimination of areas of poor potential. Such eliminations are possible by reference to the available literature alone in most cases. However, the whole process of target selection involves much more than a mere study of the literature. The following sequential approach is followed by most organizations, and is recommended.

Study of Background Literature
Before embarking on a prospecting mission, all available literature of the prospective ground should be carefully examined. Such literature may be in the form of published geological reports with maps, unpublished geological reports, memoires and records. After gathering the relevant information, the prospecting geologist should have a clear idea about the topography, geology and the ore occurrence of the prospective terrain. He will also have some idea about the climate, people, the availability of camping grounds, etc., near the prospective terrain, which is of practical importance. For locating target areas, the geologist may choose the easier and less costly ground methods when the areas are small, or the more sophisticated and also expensive aerial methods which are ideally suited for coverage of very large areas. The ground methods include reconnaissance, tracking of boulders, outcrops, etc., for evaluating direct field criteria leading to mineral discoveries.

Ground Methods—Reconnaissance
When the geologist has a sufficient number of areas which deserve field examinations, these are marked in the appropriate plans. He should now move to the field for a direct examination of these target areas. His first step should be to undertake a rapid reconnaissance of the various target areas. At this stage, no detailed examination is done, but the lay of the ground is studied carefully.

During the rapid reconnaissance, efforts should be made to see all the rock exposures. If the geological map of the area is available, this task becomes very easy. A few rapid checks should be made to study the accuracy and authenticity of the geological map also.

A prospecting operation may aim at locating one single mineral or several minerals. In either case, direct or indirect evidence of mineralization should be expected to be present in favourable terrain.

After rapid reconnaissance, the geologist will have some idea of the possible locations of mineralization. In any case, the next step is to conduct detailed field traverses to locate evidence of mineralization. Natural or man-made cuttings generally expose the rocks for easy examination. The beds of river courses, nalas and streams are the best available natural openings. If any float ore is found in such stream beds, efforts should be made to locate the original sources. This can be done by traversing the stream course upsteam. All tributary nalas, streamlets, etc., should be searched till the source is found.

Sometimes, one might suspect the presence of float ore without being able to locate it because of its small size. Such areas include the outside channel of meanders, oxbows, flood plains just after a waterfall and upstream sides of underwater obstructions. All such areas should be thoroughly searched.

Rocks and mineral deposits may be exposed in road cuttings, rail cuttings, wells, old mine workings, etc. Such areas should also be examined thoroughly. Old mine workings, ore dumps and waste dumps should be examined with great care. Any old record of the mine, plans and sections should also be examined and reasons found for stopping the mine, samples from the dumps and workings should be analyzed to see the extent of mineralization.

Various types of evidence indicate the presence of mineral deposits, depending upon the nature of mineralization, physicochemical nature of the deposit, association with host-rocks and duration and intensity of the action of weathering agents on the deposit. In areas of recent oxidation, gossans may be well preserved particularly in the case of sulphides. Such gossans may lead to ore discovery if pursued carefully. The presence of mineralization may be indicated by direct evidence in many cases.

Direct field evidence: The following features need to be examined in detail to locate mineral deposits:

1. Float ore,
2. Topography,
3. Stratigraphy,
4. Lithology,
5. Contact surfaces,
6. Structure,
7. Alteration haloes and
8. Old working, tailing dumps and evidence of smelting activity of metallic ores, etc.

Float ore: In practically every case of exposed deposits, float ores will be present. Float ores may be close to the outcrop or they may be at considerable distance. Therefore, the first thing that a prospecting geologist should look for is the presence of float ore. Float ore accumulates in certain topographically favourable terrains. Where ore is exposed in an escarpment, the escarpment will have float ore but the dip slopes may not. As mentioned earlier, float ore should be traced upstream or up the slope to locate the source rock.

Topography: All rocks, minerals and mineral deposits are consistently under attack by natural weathering agents. It is to be expected that the more resistant formations would give rise to a prominent landform. This phenomenon has presented us with two practical field guides which are in topographically prominent features and topographically subdues features. The topographically prominent features are generally associated with hard resistant ore deposits.

A typical example of this guide is the case of the iron ore deposits of India. Almost all the iron ore deposits occur on top of prominent ridges and other topographical land marks. Thus, in a geologically favourable terrain, iron ores will generally be confined to hill/ridge/plateau tops and this forms an important field guide for locating them. Iron rich gossans and cappings form prominent land marks in the Khetri copper belt in India, although it never served as a guide to the location of the deposits. Rather, the old workings seem to have acted as better guide in this case. Often subdued topography generally indicates less resistant rocks. Thus, an easily weathered rock like limestone or dolomite generally gives rise to a flat topography as typically demonstrated by the limestone.

Stratigraphy: In a sedimentary sequence, if ore is associated exclusively with a group of rocks, this can be referred to as a stratigraphic guide. A typical example is the association of iron ore with the Banded Iron formation. The association is so universal that it is safe to say that where there is an exposure of this formation, iron ore deposits may be nearby. This is clearly seen in Indian manganese ores also, but the lithological association is purely local. However, sequentially, the manganese ores occur just below the iron ores in some Indian iron ore districts. Where iron ores are known to be present, manganese ore also may be present nearby. In case of diamond deposits of India in the Vindhyan conglomerates, here, the conglomerate beds show prominent unconformity. In case of some bauxite deposits, one can also see unconformity and disconformity.

Lithology: Certain mineral deposits show specific lithological affiliation to certain rocks. This association can be recognized in syngenetic and epigenetic deposits and use in the search for minerals. Thus, chromite, platinum, etc., occur exclusively in ultrabasic rocks. Lead, zinc and silver ores show a preference for limestones. Diamonds occur only (in their original form) in eclogite (Kimberlite). Mica occurs only in pegmatites. Such guides can be used in a broad regional search and also in a purely local search for the ores. In fact, it is the primary job of the exploration geologist to establish the lighologic controls.

Contact surfaces: Contacts between two lithological units form excellent guides in the case of many deposits. This is mainly because such surfaces tend to the zones of weakness. Contacts of intrusive rocks like sills, dykes and also other intrusive, like granite, etc., are preferential targets of ore concentrations. Thus, in searching for mica, it is the pegmatite contact with the mica–schist country rock which is likely to have greater concentration of mica. Similarly, in the case of asbestos, serpentinisation and formation of asbestos are more intense near the contact of the intrusive with the limestone.

Structure: Structural features like folds, faults, shear zones, fracture zones, etc., may act as the focal point of mineralization. In such cases, the search for deposits should be concentrated around favourable structures.

Alteration holes: Mineralization associated with igneous and metamorphic processes usually leaves an alteration halo around the ore body which is direct evidence of mineralization. This halo may be defined by the occurrence of one or more hydrothermal minerals or by the intensity of fracturing or by metal ratios. Naturally, the recognition of such evidence calls for careful search, geological mapping, collection of chip samples for chemical analysis, etc.

Such haloes are common in the asbestos deposits of Cuddapah where asbestos veins are surrounded by a halo of serpentinization. Other hydrothermal ore deposits such as porphyry copper deposits and epithermal gold deposits show well defined alteration halos that can be used as exploration key.

Old workings and tailing dumps: In India where mining dates back to thousands of years, deposits which have rich outcrops are likely to have been worked at one time or the other. Old workings, tailing dumps, slag heaps, etc., are very common in the copper, lead–zinc and gold deposits of India such as Khetri copper belt, and lead–zinc belts of Rajasthan in India. The deposits of copper at Khetri, gold deposits at Kolar and other places have been found through such evidences. Tailing dumps/waste heaps may themselves be the source for some other minerals, e.g. tailing dumps at Kolar would give scheelite resources.

Aerial and special methods: Where exploration is to cover very large areas near known mineral districts, aerial and other special methods are ideally suited. Before resorting to these costly techniques, a thorough study of available literature and data of the following types need to examined and understand:

1. Aerial photographs of various types,
2. Metallogenic, mineroganic and tectonic maps,
3. Gravity, magnetic and electromagnetic survey maps,
4. Geochronologial data and
5. Geochemical and geobotanical data.

A study of the metallogenic, minerogenic and tectonic maps together with the help of aerial photographs, gravity and other anomaly maps, geochemical and geobotanical data, will make easy the selection of large target areas of exploration. In India such maps can be obtained from Geological Survey of India, the Survey of India.

Criteria for Selecting Target Areas: Presence of marketable products in reasonable quantities.

This is determined by the size and mineral content of the deposits. The size of a deposit may be highly variable under varying circumstances. It may contain a few thousand tonnes of ore or a few millions of tones. The economic viability of a deposit is often a function of its size and mineral content. If the mineral content is poor, even large tonnages may not be acceptable.

Prior to exploration, it is difficult to determine the size of the deposit. However, the geologist's experience in the district and his judgment as to the possible size of the discovery play important roles. By a process of reasoning, the relation between the size of an outcrop and the size of a deposit can be established particularly in a well-known mining district. If no such data are readily available, the nearby mines should be studied to establish this point.

Mineral deposits generally contain some portions with rich ore and some others with lean ore. In such cases, before developing or abandoning a prospect, the average grade and the corresponding tonnage should be approximately estimated. In many cases, particularly of base metals scout drill holes may become necessary at this stage.

Demand: The discovery of a deposit near an already established industry which can make use of the produced ore surely enhances its value. For example, the location of new deposits of iron ore, manganese, etc., near a steel mill will ensure their suitable exploitation. If transportation costs permit, the assurance of an established consumer at some distance can also ensure the development of a deposit quickly. Targets which fulfill the above criteria can be readily chosen for detailed exploration.

Export market: In many cases, there may not be an immediate demand for certain types of ores. If the newly located target falls within this category, then an export market should be examined. If it is found that there is a steady export demand for the material likely to be available, such a deposit can be accepted for intensive exploration.

Certain mineral based industries may need development in specific national contexts. Mining and milling of ore from such a deposit would be economically viable because of the special material demands. Under such circumstances also newly located deposits can be adopted for detailed exploration. A deposit which is near a ready market is obviously more valuable than the one located at a great distance. The market may be a plant utilizing the ore for manufacturing, or a port from where ore is exported ore, in some cases, an ore concentration facility.

A prospect can be chosen for development or rejected on the above criteria in most cases. In some cases, more complex factors are likely to come into play. For example, if a mineral is of strategic importance irrespective of any of the above considerations, the prospect may be accepted for development at the instance of the Government.

Economics of mining: Whether the ore is exposed on the surface or is exposed but continues linearly in depth will determine its amenability to being mined by cheap open-cast or costly underground methods. For underground mining to be economical, the ore should be rich in grade and should be available in reasonable quantities.

Location and transportation: The value of a deposit is also a function of the location of the deposit and the availability of a cheap mode of transportation. Topography also plays an important role. If the deposit is on a hill, transportation may be done by ropeways run on gravity which may affect savings.

The availability of readymade roads or other forms of transport adds considerably to the economic attractiveness of a deposit. In this context, the case of iron and manganese ores of Goa is noteworthy. Although generally low in grade, they have become economically viable because of the availability of navigable rivers nearby, making transportation to the shipping points cheap.

If a deposit is located near a settled township or area, labour would be available and the need for developing fresh facilities for living and recreation would be limited and thus its value would be enhanced.

3.11 Developmental Phase in Exploration

A variety of methods is available for exploration. Each has specific use, value and some inherent limitations. Choosing the techniques for specific field conditions calls for a broad knowledge of the methods and their limitations.

These methods can be broadly divided into: (i) airborne exploration methods and (ii) group exploration methods. The airborne exploration methods are listed below:

Airborne exploration method: Remote sensing Satellite imageries: Multispectral, Hyperspectral Colour

Photo-geological study: Black and white Multispectral, Airborne geophysical survey

Ground Exploration Methods: Ground exploration methods are the most important in delineating the physical boundaries are chemical nature of the individual deposits and consist of two operations, viz. detailed ground definition of targets and three-dimensional sampling. Ground exploration methods are listed below:

Detailed ground definition of targets and regional geological mapping		
Geochemical and Geobotanical sampling	Soil sampling, Stream sediment sampling, Water sampling, Rock sampling, Grid sampling of plant species	–
Geophysical methods	Gravity method, Magnetic method, Seismic method	–
–	Electrical method	Resistivity, self-potential Induced polarization, down the hole electrical
Scout drilling and rapid sampling	Three dimensional sampling Detailed geological mapping ≫ surface/subsurface	Trenching and pitting, drilling, exploratory mining

Sampling Related to the Above Operations

Out of the various methods described above, three-dimensional sampling (detailed ground exploration) is most important in the final evaluation of a deposit.

All these techniques are seldom used together and some of them (remote sensing, colour photography) are still in a developmental stage and have only experimental value except for the areas with well-developed alteration pattern. Most of the techniques listed above are available in India although airborne geophysical data are not available as yet for wide public use. Geographic information system (GIS) can be used for integration of the exploratory data both in regional and local areas. Based on the criteria defined suitable areas are delineated for further detailed exploration and drilling.

The major criterion for making a choice should be the discriminating capacity of the method being considered. Some of the techniques like regional geochemical sampling, geobotanical sampling, remote sensing, etc., have a low discriminating capacity. These techniques do not lead to a direct discovery of mineral deposits but are tools required for rather general investigations. They are best utilized when combined with methods of high discriminating capacity.

Another criterion which may be considered in choosing the technique is the status of the terrain. In virgin areas, one might have to choose large target areas and it is important that the first stage of elimination be done rapidly. Here, choice may be made in favour of rapid technique even if they have poor discriminating capabilities. Expense and ready availability of equipment and skills also may influence the choice of techniques.

3.12 Ground Exploration Methods

By aerial or other methods, target areas are selected for ground exploration. During ground exploration, the deposits will have to be broadly delineated. Such delineation can be achieved by regional geological mapping, scout drilling, followed ultimately by three-dimensional sampling.

Detailed Ground Exploration of Selected Areas

3.12.1 Regional Geological Mapping

It is done for delineating potential mineral bearing areas, and is generally done on a small scale topographic sheet of scale 1:250,000–1:50,000. Such maps are not made by exploration company, since maps are often available from geological surveys of the country concern. If such maps are not available and the area is totally virgin then only the exploration agency need make maps.

An exploration company shall delineate potential mineral bearing areas on the basis of rapid field surveys. Outcrops, river and stream cuttings, old workings if any are utilized to mark outpromising areas may be chosen and a more detailed map on a scale 1:25,000 and 1:5000 indicating the mineral deposit, outcrops and such details like workings.

The geological map aims to show the geological set up of an area in terms of the occurrence and distribution of rock types, their mutual relationships, attitudes, structure, association with topography and drainage.

Since geological map is done by collecting and recording geological data on a base map, the first need is a base map. This base map gives the permanent topographical and drainage details of the area and also other man-made details such as roads, railways, villages, etc., the maps is suitably contoured with reference to the permanent bench marks and standard co-ordinates. In case of large-scale plans, a system of artificial coordinates is applied for local reference value. For geological mapping on a regional scale, small-scale topographical sheets are good enough, cadastral maps are used for some large-scale work. But for large-scale work, a base map according to need is prepared.

A geological map can be generated by a ground survey or on serial photographs. In ground survey, geological feature is mapped by a compass and tape, or pacing traverses from known fixed points indicated on the map. In case, more precision is required, a theodolite is used. Plane table, telescopic alidade, are also be used where detailed mapping is needed.

3.12.2 Geological Mapping by Compass and Tape/Pacing

The method is applicable in the case of mapping on small-scale toposheets or on large-scale maps. In either case, the principles involved are the same and can be summarized in the following steps.

(i) Orientation of the map: This is done by finding the magnetic north of the place by compass and aligning the north of the map absolutely parallel to the compass needle.

(ii) Location of oneself on the map: Objects or landmarks on the ground which are indicated on the map are identified. A bearing to these objects is taken. The distance is estimated either by tape or by pacing. The angle and distance are plotted and the position fixed. In case it is possible, the position can be fixed by taking a bearing each form two objects and by locating their plotted intersection.

(iii) Mapping: Various geological data can be systematically plotted by the methods mentioned above. Various geological data are located by a single bearing and measuring of distance or by intersection of two bearings from various fixed landmarks. Where actual measurements are involved, the distance can be measured by pacing or by actual tape measurements.

Geological Mapping by Plane Table

This method is very useful in large-scale geological mapping, particularly when the ground is not rugged. A plane table consists of a drawing board fitted on a tripod in such a way as to enable levelling and orientation. The base map is fixed on the board firmly and an alidade, occasionally with telescope and vernier, is used for taking the bearings. Instead of an alidade, open sights fixed on a straight edge are also used. In mapping with a plane table, first, two known points located on the base plans are identified on the ground. A base line is fixed between these points which are designated as A and B. To locate the points to be mapped, first, the plane table is fixed directly above the point A and the table aligned with the alidade so that the base line AB is in direct line with point B. Now, the alidade is turned toward the point to be mapped, say C. After aligning the alidade, a ray is drawn from A in the direction of C. Now we have the ray AC. The table is shifted to B and the base line BA is aligned with point A as was done for AB. Now the alidade is pointed toward C and the ray BC drawn. The two rays will intersect at the position C which is the map position of the object C.

Plotting can be done by direct alignment of the alidade with an object of interest and measuring the ground distance by tape. Here, only one ray is involved and the distance is actually measured and plotted.

Geological Mapping by Theodolite

To handle a theodolite, it is always preferable for a professional surveyor to join the geologist. The principals involved here are not very different from those described for compass traverses and plane table survey. The major difference is that a theodolite can be used for tachometric survey for finding the distance between points. This is a great advantage where linear tape measurements are difficult due to terrain conditions. For mapping in open pit or underground, theodolite is the best instrument. When a surveyor and geologist work together, the work of mapping and surveying can be combined. This will effect considerable savings in time.

A geological map would show at the following features:

1. contacts between various formations (formational in small scale maps and lithological in large-scale maps),
2. strike and dip of contacts,
3. outcrop position of faults, folds, etc., with strike, dip and plunge where observable,
4. planar features like cleavage, schistosity and joints and
5. soils and geomorphic features which have interpretational geological value. These may include alluvial formations, landslides, meander bends, etc.

In large-scale maps, a amount of details will be large and even very small-scale features might have to be mapped in. A mine plan for example might show even the position of thin stringers and veins and their lithological and structural disposition. Anything of interpretational value should find a place in a geological map.

The reading and interpreting, the two-dimensional data depicted in a geological map will have to be imagined in a three-dimensional plane. A projected picture of the structure below the ground is very essential. This is obtained by the construction of cross-sections. Here, first a line of section will have to be chosen. The maximum information is generally available in a section line which is across the general strike of the formations. A profile is drawn by placing the contour values on the section line and projecting from the lowest available elevation. The various elevations are then projected by scale from the base. Next, the various geological and lithological contacts, fault traces, fold traces, etc., are plotted on the profile outline showing their respective dips, in amount and direction, and appropriate projections made to bring out the sub-surface disposition of the deposit. The resulting picture provides a fairly good view of the geological features in the third dimension.

In the geological map, it is customary to show various geological formations in symbols or in colour.

Boulder Tracking or Float Ore Tracing

Float ores are boundary ore fragments period out of outcrops of mineral deposits by the process of atmospheric weathering. These fragments are distributed by surface drainage and wind over large areas by transport agencies like river systems and other weathering agents. Naturally, they are found in stream beds and stream banks more frequently. They may occur in alluvial formations also. Generally, the angular edges of a float indicate the nearness to the source in contrast to rounded pebbles which signify considerable transportation over a long distance. The float ores should be traced upstream to locate the original outcrop. This can be done by systematically traversing upstream and examining every water course to its origin.

The minerals like gold, diamond, cassiterite, etc., panning is advisable to locate minute quantities of the ore as explained earlier. It should be done systematically, also upstream as in the case of float ores. Panning can accomplish two tasks, viz., the location of areas of eluvium and alluvium carrying increasing quantities of minerals in the form of placer deposits.

3.13 Geophysical Methods

Geophysical prospecting is the search for hidden mineral deposits by the measurement of certain physical properties of the earth's surface. This involves the field measurement of certain properties like gravity, magnetism, seismicity and electrical conductivity of rocks by sensitive instruments and the interpretation of such data along with the available geological data. The knowledge of the potential of this method is essential for exploration geologists. Five major geophysical prospecting methods are available: these are gravity, magnetic, electrical, seismic and radioactive methods. Since this bulletin deals with non-radioactive mineral deposits, only the first four methods are discussed here. Despite a high degree of sophistication in instrumentation and skill, mineral discoveries by this method have been relatively few even in industrially advanced countries.

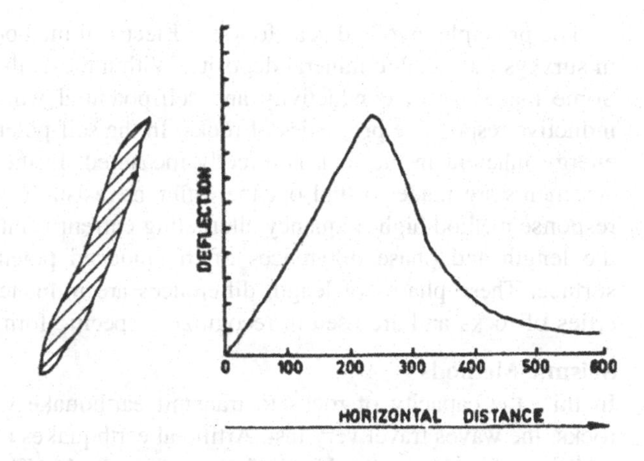

Fig. 3.1 Sulphide ore body and gravity anomaly curve

Gravity Method

The principle utilized here is to locate the anomalies of gravity from place to place in an area of search. If a plumb bob is suspended freely it will point towards the centre of the earth. If a heavy mass is nearby it will exert a gravitational influence on the plumb bob with the result that the bob will show a deflection. The heavy mass may be a body of massive sulphide ore, iron ore, bodies of chromite, etc., all of which are known to cause such deflections. The deflections are measured by instruments. Three such instruments are in common use. They are the pendulum, torsion balance and gravimeter. Of these, the gravimeter is the most useful.

In the field, traverses are laid at intervals, varying with the scale of operations, just as in the case of geological traverses. Measurements of gravity are made at predetermined points and the readings are plotted on a graph with distance on the *X*-axis and deflections on the *Y*-axis.

The resulting graph will show an "anomaly" if the traverses is over a body which exerts a gravitational pull of its own. Anomalies can be interpreted by contouring also where the peaked values will be better reflected in the areas of anomaly (Fig. 3.1).

Magnetic Method

Certain rocks have magnetic properties. They exert their magnetic influence above the normal magnetic field of the earth. Certain minerals which contain magnetite in a subordinate form such as magnetite in asbestos, and pyrrhotite associated with base metals also may exert magnetic influence. This magnetic influence causes deflections of a magnetic needle which are measured by sensitive instruments such as a magnetometer on traverses just as in the case of gravity measurement. The data can be interpreted in the same way as gravity data.

Electrical Methods

The following methods are generally recognized: resistivity, self-potential, induced polarization and down the whole electrical methods.

The principle involved is as follows: Electrical methods are particularly relevant in surveys for metallic mineral deposits. Within these, there are various techniques. Some make use of conductivity and self-potential while others make use of the inductive responsive properties of rocks. In the self-potential method, the electrical energy inherent in the rock is directly measured. In the resistivity methods, measurements are made to find out the different resistivity of rocks. In the inductive response method high frequency alternating current is introduced into the earth and the length and phase differences of the induced potential are measured on the surface. These phases of length differences are influenced by the electrical properties of rocks and are used in recognizing specific formations.

Seismic Methods
In this, the capacity of rocks to transmit earthquake waves is studied. In dense rocks, the waves travel very fast. Artificial earthquakes are produced by explosions and the waves measured by sensitive seismographs. There are two types of measurements in this method. One measures the property of reflection and the other the property of refraction. In the reflection method, the time taken by the various surfaces to reflect seismic waves is measured. In the refraction method, the principle used is that surface waves travel at a slower speed compared to the deeper waves.

As we have seen during the course by this handbook, the process of mineral exploration and evolution could be divided into:

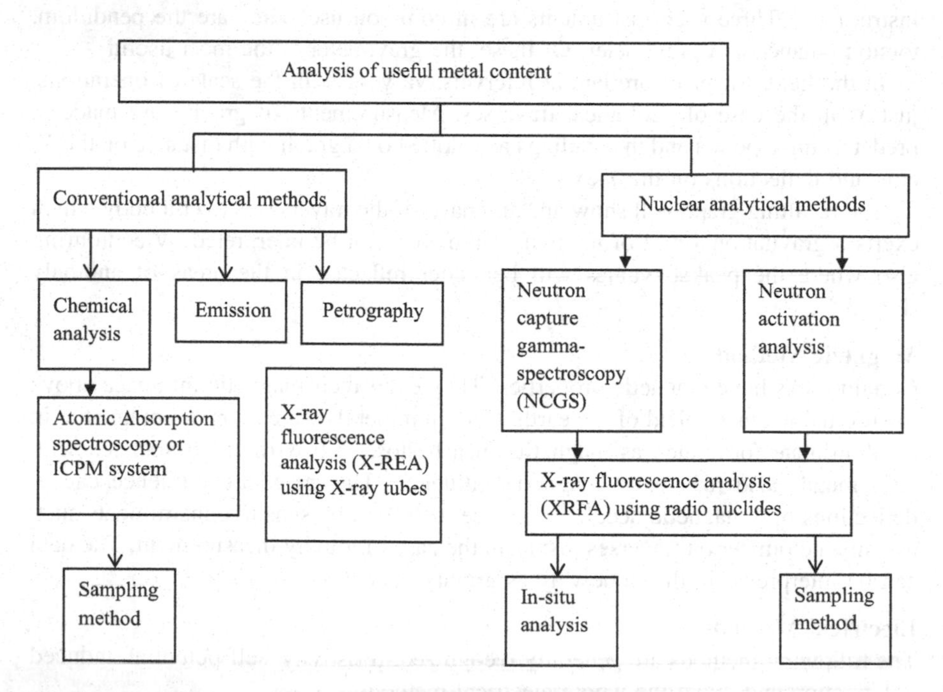

Fig. 3.2 Analytical methods as their applications

(A) general reconnaissance or primary study
(B) prospecting and feasibility study

3.14 Nuclear Methods

It is in exploration stage that nuclear methods can be used along were conventional procedures. The content of useful components is an important factor in determining mineability of a deposit. Nuclear or, more accurately, game spectroscopic method of analyses, is element specific and in conjunction with on-line data processing they permit ore concentration to determine with a minimum delay, as given in Fig. 3.2.

The conventional analytical procedures make no claim to being complete. In general, however, in the case of conventional methods the analysis is carried out on soil and rock samples obtained from drill cores, whereas rock gamma spectroscopic procedures permit in situ analysis and this offer a number of distinct economic advantages.

- Some of the expensive core drillings can be replaced by cheaper full hole drillings.
- Direct analysis into bore hole replaces such activity as classifying, transporting, drilling and processing drill core.
- Volume of rock covered by the borehole analysis is generally longer than the drill core volume. This makes the analysis more accurate.
- The analytical results are immediately available.

Some of the system of nuclear method for exploration is listed.

1. Gamma spectroscopic analytical procedures.
2. Neutron captures Gamma spectroscopy.
3. Neutron activated analysis.
4. X-ray fluorescence analyze.

However, the cost-benefit analysis by the methods is to be evaluated prior to adopting these methods.

In the deep-sea mineral exploration for manganese nodules radioisotope EDX system was successful demonstrated (Kunzendorf et al. 1974).

All the methods, except the seismic, are capable of being executed by airborne instruments for covering large areas rapidly. Electrical methods offer the best scope for application in small areas.

Geophysical methods are best done by geophysicists. The exploration geologist's function is essentially to decide at what stage and in which terrain geophysical surveys are most useful and then to carry out the follow up action on the basis of geophysical findings.

3.15 Geochemical Mapping and Sampling

The quantitative determination of the anomalous distribution of certain elements in the surface soil and sub-surface overburden and comparison with a known background value give an idea of the rocks and mineral deposits hidden below the ground. This is the essential principle of geochemical prospecting.

The common practice followed in geochemical method is the collection of soil, stream-sediment/water/rock chip samples at regular intervals, the grid lines being transverse to the strike of the suspected zone of mineralisation. Air samples are also collected for certain geochemical surveys. The metal content is determined by rapid geochemical analysis of samples. Geochemical methods are very effective in locating targets of basemetal mineralisation and are used extensively in India.

Soil sampling: Residual soils are common in both temperate and tropical regions the latter giving rise to very thick soil caps. These soils always carry mineral particles of the parent rocks. Soil geochemical sampling aims at recognizing such values, anomalously high value concentrations indicating a soil-capped target. Samples of 100 gm, are collected at predetermined intervals for rapid analysis.

Stream sediment sampling: This is the most widely used technique in geochemical surveys. Fine stream sediments are collected from surface stream channels. These sediments usually carry mineral matter in fine form derived from their original source rock. Samples are collected from the active stream channels or from flood plains. The samples should be collected at intervals of 0.15 km.

Water sampling: Water sampling is of relatively little value except in the case of uranium and, in rare cases, zinc and copper. Water samples are collected from lakes, rivers, and also from underground sources.

Rock sampling: Rock samples are collected from visible rock outcrops by cutting out chips of rocks from predetermined grid points. Samples of—100 mesh size are prepared for analysis.

One of the major problems in geochemical exploration is the treatment and interpretation of data. A single investigation may produce some 50,000–60,000 individual readings and they have to be classified, processed and analyzed to produce any useful information. Generally, such studies require the help of a computer, or at least an electronic desk calculator.

In presenting the interpreted data, one which is easiest to understand is the method of value contouring. Geochemical prospecting data can be more effectively interpreted by following graphical statistical procedures or the computer methods. Briefly the methodology is based on the following concepts. Any area under investigation will have three types of geochemical values. One is the background value which describes the average metal value of the area. The other is the threshold value which describes the zone separating the average values from the anomalous values. Lastly, there are the anomalous values which are above the threshold values. The area of anomalous values forms the best target for exploration.

Table 3.21 Leaf characteristics

	Element or mineral	Mutational effect
1.	Chromium	Chlorosis of leaves
2.	Cobalt	Increase of chlorophyll in some species and chlorosis in others
3.	Copper	Chlorosis of leaves and dwarfism
4.	Iron	Darkening of leaves
5.	Manganese	Chlorosis of leaves with white blotches
6.	Molybdenum	Formation of abnormally coloured shoot
7.	Nickel	Chlorosis and necrosis of leaves
8.	Serpentine	Dwarfism; colour changes of flowers
9.	Uranium and radioactivity	Variation in flower colour, presence of abnormal fruits, increase in chromosomes of nucleus, stimulation
10.	Zinc	Chlorosis of leaves; symptoms of manganese deficiency

Geobotanical Mapping and Sampling: The presence and variety of a particular plant species in the area of mineralisation have been recognized as a guide to locating ore since many years. However, it is only in the recent times that it has started becoming an organized science and a method of exploration.

The growth and distribution of a particular plant in a given area are known to be influenced by sub-surface geology. Geobotanical survey involves a visual study of the nature and distribution of the vegetative cover, plant distribution, the presence of indicator plants, mutational or morphological changes in species introduced by mineral enrichment.

The mapping technique in geobotanical surveys is as follows. Plots of 5 km^2 are chosen and the plant species found in them, their growth, density of growth, new species rare species, etc., are marked out. A species which shows special affinity to the area is chosen by a process of progressive elimination. The association of these species with any known mineral occurrence in the area is established if such a correlation exists. This correlation can be used for searching adjacent larger areas.

However, tracking of indicator plants is more useful in this type of exploration. Two types of indicator plants are recognized. These are—(i) universal indicators and (ii) local indicators. Universal indicators grow only in a mineralised terrain and are seen the world over in similar setups. They are of course rare in their occurrence and distribution. Some of the common universal and local indicators are given in Table 3.20. The list also shows the effect of the presence of certain metals on certain plant species. These also may form useful guides.

The present of these plant species is indicative of the presence of the corresponding elements. Some mutational effects caused by the presence of certain elements on certain plants are also very useful in recognizing them. This recognition is particularly significant in the case of multispectral aerial photography where the mutational effects can be recognized. Thus, they provide an indirect evidence of the possible presence of certain elements in regional scale explorations.

A list showing the mutational effect and the corresponding causative elements is Table 3.21 more common. Some 63 species of local indicators have been recognized. But opinion is divided about their ultimate practical utility.

3.16 Other Field Procedures

Scout drilling and rapid sampling: Scout drilling is done exclusively for confirming the presence of deposits discovered on the basis of exploration conducted by geochemical, geobotanical and geophysical methods. Scout drilling can be done by coring or non-coring drills. The samples resulting from such drilling are analyzed to study the major elements and their percentage availability by rapid methods. Scout drilling helps in understanding the nature of a newly discovered deposit.

Three-Dimensional Sampling: Three-dimensional sampling aims at establishing the dimensions and grade of the deposit in all its details. This operation involves detailed geological mapping, trenching and pitting, drilling, exploratory mining and sampling pertaining to each operation.

Detailed geological mapping: The methods of mapping have already been discussed. The detailed geological mapping, the details plotted are substantially more and the scales may be of the order of 1:2000, 1:1000, 1:500 or even larger with contours of 2–5 metre intervals. Closely-spaced grid points should be made use of in actual plotting. The details to be plotted may be lithological boundaries, ore type contacts and structural features. Detailed mapping should aim at establishing correct boundaries of all these features and should help in depicting the correct two-dimensional picture of the deposit.

Detailed mapping should also be done to project subsurface information which should be confirmed by drilling, and thus will form the basis of planning out a drilling, pitting or trenching programme. When a deposit is being proved by exploratory mining, underground mapping of the various openings may become necessary. Methods of underground mapping are discussed below.

Underground Mapping: Underground mapping consists of mapping of levels, drives, raises, winzes, crosscuts, stopes. Although the broad principles of geological mapping discussed earlier are valid for underground mapping, the latter requires certain special skills and also a slightly different methodology. Particular attention has to be paid to the plane of projection, which is a horizontal plane. In underground openings, all mineable details will have to be visualized in three-dimensions. In order to depict this three-dimensional view, all mapping data should be projected on to a single plane. In most cases, this plane coincides with breast or waist level. In certain special cases, the plane may coincide with the back of the drift.

The underground map base should be an accurately prepared survey plan showing all details of underground working. In a working mine, such up-to-date maps are always available. The map should be of a large scale (1:200, 1:100 or in some vein and stringer type of deposits even larger scales may be necessary). Since underground workings are often dirty, maps and instruments used in mapping should be properly protected by sturdy waterproof material.

In underground mapping, the most versatile tool is the Brunton Compass for taking bearings and dips and strike of linear features. For taking dips underground, under conditions of poor light and awkward points, a clinometer compass is more suitable. However, the normal practice should be to establish a large number of stations, evenly distributed in every working, which can act as reference points from which measurements may be taken. Such reference points should be surveyed by theodolite and connected to the main survey. The stations should be marked by luminescent paint for visibility.

Various items of geological significance to be mapped are summarized below.

lithology of the various rock units exposed with their contacts, ore zone with its footwall and hanging wall contacts ore types and various host rocks of each type of ore where feasible, strike and dip of the various formations and also various planar features like cleavage, joints, contacts, etc., structural feature like folds, faults, unconformities, intrusive rocks, crushed zones, ore solution channels, etc., wherever possible, and reaction rims, zoning, etc., Plotting of all details should invariably be done simultaneously with the measurements.

Special care and attention are required for measuring the various planar features underground. In taking strike, the planar features should be taken, with the reader standing close to the exposure on one wall and aligning the compass exactly in line with the exposure of the same feature on the other wall. Measurements from known points nearby on both walls can be of help in confirming measured positions of the planar features on the map[7]. In most cases, the wall exposures of the planar features may not be clear enough to take a direct measurement of the dip. In such cases, the procedure is to measure the correct strike and then measure dip exactly at right angles, with the clinometer plane, coinciding with the exposed dip plane. This can be done very well by holding the compass in hand and aligning the two planes by eye. Direct measurement of dip should be resorted to only when the dip plane is very clearly exposed on one wall.

In underground measurements of strike and dip, the help of a string is recommended. By tying the ends of the string on both walls of the opening suitably, dip and strike measurements become easy. In dip measurements, the string is aligned to the plane of dip by tying it at two points within the dip plane. Aligning the compass to the string may prove easier than aligning the compass directly. Dips may be measured by a plumb bob too. For this purpose, a plumb bob is suspended from the top of the exposure. The distance from the plumb bob to the back of the planar feature is then measured. The length of the plumb line divided by the distance of the plumb line to the back of the planar feature gives the tangent of the angle of dip.

Trenching and Pitting: A trench is a narrow linear excavation which is generally done to expose concealed outcrops where the soil or other secondary cover is not too deep (say less than 1 m). A trench may be very deep and very long in certain

Table 3.22 Presentation of trench data—suggested guideline

<table>
<tr><td colspan="5">Name of the Investigation</td></tr>
<tr><td colspan="2">Trench No.</td><td colspan="3">Date of commencement</td></tr>
<tr><td colspan="2">Location in co-ordinates</td><td colspan="3">Data of completion</td></tr>
<tr><td colspan="2">Length</td><td colspan="3">R.L. of floor and top of trench</td></tr>
<tr><td colspan="2">Width</td><td colspan="3"></td></tr>
<tr><td colspan="2">Depth</td><td colspan="3"></td></tr>
<tr><td colspan="2">Recorded by</td><td colspan="3"></td></tr>
<tr><td colspan="2"></td><td colspan="3">Log of each wall</td></tr>
<tr><td>From Footwall</td><td>To Hanging wall</td><td>Lithology</td><td>Contact dip and strike</td><td>Planar strike and dip</td></tr>
<tr><td></td><td></td><td></td><td></td><td></td></tr>
</table>

Other Details:

(1) Total volume of material excavated in m^3 estimate of total weight of material excavated
(2) Number of samples taken with details
(3) Thickness of overburden if any

specific cases where the distribution, size and shape of the mineral concentration are to be studied. Normally, however, trenches are put to expose the outcrops preferably from the footwall contact to the hanging wall contact. The actual excavation may be done either by manual methods or by excavating machines.

Trenches are most valuable when the ore outcrop is narrow and linear. In a capping type of deposit where the outcrops have larger widths, trenches may be hundreds of metres long and may be too expensive. Similarly, wherever excavations are deeper than one metre trenching may become very expensive. Briefly, therefore, trenching may be chosen as an exploration method only when the outcrops are narrow and the cover thin.

Trench may be done for purely exploratory purposes such as exposing the outcrops, sampling, etc., or for purpose of mining in which case it serves a dual purpose, that of exposing the ore body and later for developing the mine. A trench of the latter type will be of larger dimensions but generally pays for itself through the ore produced in the process.

A trench provides considerable geological information, the recording and safekeeping of which will be of great importance. Table 3.22 gives a guideline which may be used for the collection of trench data. Trenching has been adopted as a major exploration method in many iron ore and bauxite deposits. Trenches can give detailed information about the nature of the hanging wall and footwall rocks, the lighological and grade variation of the ore body, the structural details of the ore body, etc.

Depending on the topography and strike continuity, an ore body maybe studied in a number of trenches at close intervals. The information from close by trenches can be compared and the strike continuity of the ore body established by interpretation. Similarly, if trenches are at various elevations of the same ore body

Table 3.23 Presentation of pit data—suggested guideline

Name of the Investigation	**Date of commencement**
Pit No.	Date of completion
Location in co-ordinate	
Pit top measurements (a) Length (b) Breadth	R.L. of pit top R.L. at pit bottom
Pit bottom measurement (a) Length (b) Breadth (c) Depth	
Branches if any with their measurements	
Recorded by :	

Face I (give for all the 4 faces)

From	To	Lithology	Contact dip and strike	Planar strike and dip

certain depth wise predictions be made. Wherever possible, the trench should be cut close to and parallel to the nearest dip profile section for easy interpretation. When surface trenching is followed by boreholes, the data of the corresponding pair of nearest borehole and trench will be of great help in studying the variability of the ore body at depth.

Sampling in Trenches: In sampling a trench, it is better to cut channels, restricting channel lengths to specific types of ores or group of exposures, depending on the type of mineralisation. Details of which are given in Table 3.22.

To be accompanied by a sketch map prepared on scale 1 cm = 1 m or 1 cm = 2 m showing all details.

Ore material dug out of a trench as a whole can also be used as a sample for bulk test. It is seldom used for chemical analysis due to the labour involved in breaking down to the proper sample sizes. In case the outcrop is weathered, the trench should go deep enough to expose fresh ore.

Pitting: Pitting is the process of digging openings to penetrate soil cover and other loose material to reach mineral ore bodies concealed underneath. Pitting is also done for the specific purpose of sampling an already existing ore body or for testing the depth or thickness of exposed ore bodies and outcrops. A depth of penetration up to 30 m. is possible by pits under ideal conditions. But pits of over 12 m are seldom feasible and even if technically feasible is likely to be prohibitively expensive.

The major advantage of a pit is that the opening is sufficiently large so that the exposed section of the ore body can be directly examined and if necessary sampled repeatedly. For collecting and compiling the pit data, a guideline is given in Table 3.23 and Fig. 3.3.

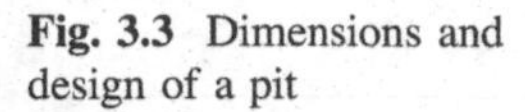

Fig. 3.3 Dimensions and design of a pit

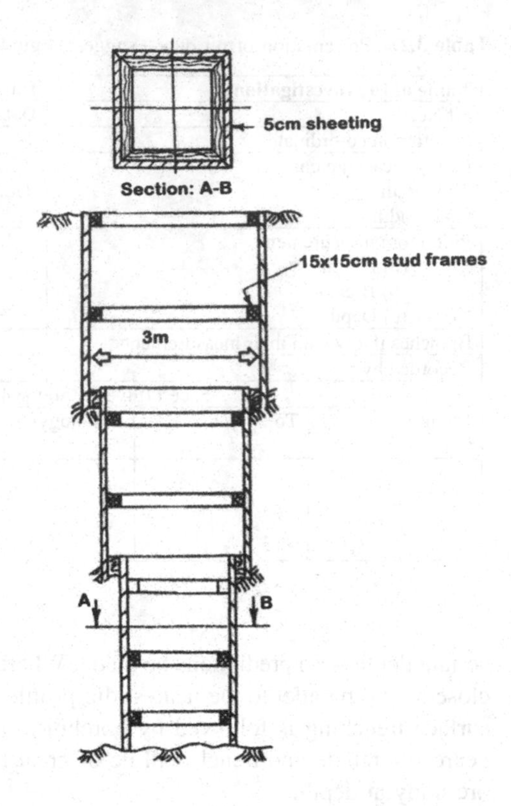

Pitting can be done manually or with the help of machines. In many cases, when the ore is hard, some amount of drilling and blasting might become necessary.

Pitting is a very widely used exploration method and may be very effective in proving deposits which are widely exposed on the surface and are relatively thin. They are least effective when the outcrop is linear and narrow and the ore body steeply dipping. Pits may be sunk for purposes of developing them into a mine in which case they serve the purpose of exploration as well as exploitation. Such pits are referred to as trial pits and may have a very large opening. In opening such pits, one is assured of a return in terms of the mined ore, the sale value of which may underwrite the cost of pitting. Such considerations are often valid in trenching also. Pitting has been exclusively used as an exploration method in many bauxite and iron ore deposits.

Sampling of Pits: Pits usually expose ore body on the four walls. It is not necessary in all cases to cut channels in all the walls, particularly in mineral deposits like iron ore, limestone, etc. But in deposits like bauxite and manganese, where the ore is lensoid and the ore values tend to be erratic, it is advisable to sample all the four walls and make a composite sample or have four separate samples analyzed and the values combined by weighted average.

The whole material dug out from a pit car also forms a sample but as in the case of trenches, bulk reduction.

Other details

(1) Volume of the final pit
(2) Volume of the total excavated material
(3) Volume of the ore portion
(4) Volume of the reject (undersize, undergrade, etc.)
(5) Volume of the overburden and details
(6) Number of samples from the excavated material
(7) Number of details of channel samples

All details need to be accompanied by sketch map of each wall of the pit on scale 1 cm = 1 m. is laborious and may be avoided. In float ore exploration, however, the whole material has to be treated as a sample. In order to know the depth-wise distribution of ore every 50 cm or 1 m depth of excavation may be separately sampled. The ore recovered is also subjected to screening to study size.

3.17 Drilling

Drilling or boring is the process of driving holes into rocks. The material which is cut during the process of drilling may be used for purposes of tests, chemical analysis, etc., which is the prime purpose of drilling in mineral exploration. Drilling may be to very shallow depths in which case simple wash boring can be made use of. In the case of mineral exploration, depths of hundreds of metres will have to be penetrated which require the use of sophisticated equipment.

Five methods of drilling are generally recognized. They are: percussive, attractive, rotative cutting, rotative shearing and rotative crushing.

In percussive drilling, the rock is broken by repetitive impaction. In this group, five types of drills can be recognized. They are pneumatic rock drills, down-the-hole drills, independent rotation drills, motor drills and cable-churn-drills. In attractive drilling, the rock is ground away by abrasive action. Two types of drills are recognized in this group. They are diamond drills and shot calyx drills. In rotative cutting drilling, the rock is cut or planed away. All auger drills fall within this group. In the rotative shearing method, rocks are fragmented by wedge action. This group includes drag-bit drills and rotary percussive drills. In rotary percussive drilling also, rocks are broken by wedge action, but the wedge action is used in combination with a thrust. All heavy rotary drills fall within this group28. Of all the drilling methods and drills discussed above, diamond drilling is the most versatile and of maximum use in mineral exploration. Churn drilling and rotary drilling also find use in mineral exploration although their use is rather limited.

Any drilling programme should aim at establishing the strike-wise and dipwise continuity or discontinuity. For this purpose, the interval of drilling has to be chosen on the basis of available geological information. This should be kept in view for establishing the continuity of the ore body at depth and also for intersecting the ore body at various future mining levels. Drilling also should aim at establishing the various structural projections.

Diamond Drilling: Diamond drills can be used for surface as well as sub-surface drilling30. In diamond drilling, a cylindrical bit (cutting tool) impregnated with diamonds is connected to a string of hallow jointed tubes and rotated by a mechanical device which may be a diesel engine or a pneumatic compressor device. Water, drilling mud, and in some cases air is circulated through the hollow tubes to keep the bit cool and wash out the cuttings. The diamond cutting edge cuts a cylindrical core as it penetrated the various rock strata. Such a core is collected in a core barrel which is fitted immediately above the bit. From the core barrel the core is collected periodically either by withdrawing the tool string (conventional method) or by removing only the barrel and bit (wireline core barrel). In drilling shallow or moderately deep holes, the time consumed in lowering the drill bit, core barrel, etc., into the hole and hoisting them up repeatedly for collecting the core may be less significant than other considerations. Where time is an important factor, wireline drilling may be resorted to. Wireline drilling differs from conventional drilling in that the inner tube that contains the core can be lifted to the surface without hoisting the entire drill string, by releasing it from the outer tube of the core barrel. The core so collected forms the sample which may be subjected to tests or analysis. The core barrel is constructed in such a way that core is retained during rotation. To ensure complete recovery, the core barrel may have an inner stationary tube which retains the core with an outer rotating tube which transmits the rotation to the bits. Such a core barrel is called a double tube core barrel. In a double tube core barrel, water is circulated between the outer and inner walls and the core is not washed by water; because of this, better core recovery is ensured, particularly in weak and fractured zones. Drilling can also be done without the circulation of water in which case it is called dry drilling. In dry drilling, the core recovery is always very high, but the bit losses are also correspondingly high. Table 3.5 shows the sizes of bits and barrels.

Calyx drilling which is of fairly wide application is done on the principle of diamond drilling but instead of diamond bits it uses chilled steel shots as a cutting tool.

These days, however, as a substitute to diamond drilling, dry drilling by tungsten carbide bits is being resorted to quite extensively by many exploration agencies.

Churn Drilling: In churn drilling, a hollow tube suspended on a cable and attached with a cutting tool is driven into the ground by rectilinear motion. The motion is imparted to the bit by continuous dropping and rising of the tool string. The cuttings so generated are collected periodically by sludge collector or bailer. The method is not widely used in mineral exploration. However, in the exploration in alluvial terrains, this type of drill is excellent.

Rotary Drilling: In cases where the information sought is purely of confirmative nature and can be interpreted from chips and cuttings, rotary drilling is useful. In mineral exploration, such drilling can be used most effectively in the mine exploration stage when bench drilling becomes necessary.

Apart from the drilling methods described above, some simple devices like augur drilling are also used while exploring for soft and non-compact material like clays and bauxite.

Table 3.24 Presentation of Drill Hole Data—suggested guideline

(a)

Name of the Investigation	:	Date of commencement :	
Borehole No.	:	Date of completion :	
Angle of inclination	:	R.L. of the collar of the borehole:	
Bearing	:	Total depth of borehole :	
Location in co-ordinates	:		
Recorded by	:		

(b)

Sample no.	Length of run		Total width	Recovered length / weight of core/sludge recovery	Core size	Sampled section (m)	Rock type and lithological description	Length of sampled section	Extra-polated length of sampled section	Specific gravity	Assay X	Width X assay	Sampled width	Average of ore zones		Nowen-clature of zones	remarks
	From	To												True width	Average assay		
(1)	(2)	(3)	(4)	(5)	(6)	(7)	(8)	(9)	(10)	(11)	(12)	(13)	(14)	(15)	(16)	(17)	(18)

Collection and Preservation of Drill Samples: Three types of samples come out of drilling; core samples from diamond drills or various other coring and calyx drills. Wet cutting or sludge from churn drill, diamond drill, or wet rotary or percussive drilling; and Dry cutting from air-flushed diamond rotary augur or percussion drilling.

Among all these, the diamond drill core and the calyx drill core can be considered almost identical with the host rock from which they have been removed. When the core recovery is 100%, the core reflects the strata faithfully. However, full core recovery is seldom achieved. When core losses occur, the sludge (cuttings), coming out with the circulating water mud or air must be given due weightage. In other forms of drilling, only sludge is obtained, Table 3.6 shows the core, sludge ratios and their computation.

In core drilling, the core is the principal sample and is collected from the barrel at intervals of 3 or 6 m. Shorter drill runs and shorter core lengths are preferred in some cases when specific information is to be had from short lengths of core. After recovery from the barrel, the core is dried to remove only the superficial moisture and kept in a core box. For preserving and indexing the drill cores, the Indian Standards Institution has evolved a standard procedure IS-4078: 1967 which may be referred to for proper upkeep of drill cores. The method is demonstrated in Table 3.24a, b. The core box may be made of wood or aluminium and should be sturdy and durable.

Portions of core which are needed for analysis are first split into two by a core splitter. One portion is kept in the box. The other portion is again split into two. One portion is sent for assaying. The other is sent for physical and mineralogical studies. When the core recovery is poor, the sludge representing the core loss will have to be assayed and computed with the core portion.

Sampling of Drill Core, Sludge and Cuttings: It was mentioned earlier that the core sample is split and one is taken for analysis and another kept for reference. That half selected for sampling is again split into two, longitudinally and one half is taken for sampling. Sampling may be made from 50 cm, 1 m, 2 m, or even 3 m section lengths of core. Or even the whole mineralized band can be taken as one sample, if the mineralization is visibly of uniform quality and distribution. When mineralization is irregular, shorter lengths of samples are better. However, less than 50 m, length samples are impractical in most cases. The selected core lengths are sized down to the appropriate sample size and sent for analysis.

In core drilling, as mentioned earlier, it is rarely that recoveries are 100%. For computing the loss of core the Table 3.24a, b given earlier should be helpful. The table shows core to sludge ratios in core drill samples. It becomes necessary to compensate for this loss. This is done by sludge analysis. However, the sludge and core assays cannot be combined by weighted averaging. One easy method is suggested below.

$$A = \frac{C}{L} \times \frac{D_1^2}{D^2}(A_1 - A_2) + A_2$$

where

A average assay
C recovered core
L Length of the hole (relevant to the calculated portion)
D diameter of the hole
D_1 diameter of the core
A_1 core assay and
A_2 sludge assay

In the case of non-coring drilling, sludge or cutting is sampled. Here also, analysis may be of samples of 50, 1, 2 or 3 m lengths depending on the type of mineralization.

Deviation and Surveying of Drill Holes: Drill holes may be vertical or inclined. When inclined, it is usual to plan the intersection angle of the tool string with the hanging wall contact of the projected ore body at about 90°. In any case, the exact angle at which the tool string is aimed to go through is never achieved. An angle is often made either with the vertical plane of the hole or the horizontal or even at angles to both the planes. Although basically a drilling problem, an exploration geologist should be conversant with the phenomenon of hole deviation and the various devices available for correctly surveying these deviations, to enable him to make correct sub-surface interpretations. Most of the deviations are caused by a sudden change in the hardness of various drilled formations, fracture zones, joint planes, etc., present in the formations. Deviations can also be caused by poor equipment and improper alignment of the machinery. Deviations pose serious problems in deep drilling, particularly directional drilling with a small angle from the horizontal. Since the deviation of a borehole from its assigned course is very common, some corrective steps have been developed for either bringing back the hole into the proper course or hitting the formation at a known point. This method is called directional drilling. It can be achieved by employing some special type of wedges. One side of the wedge is firmly anchored to the wall of the hole and the other side that projects out of the wall helps the drill bit the man oeuvre into the desired course.

An accurate measurement of the angle at close depth intervals and direction of deviation is essential for a proper interpretation of subsurface information. This measurement is called borehole surveying. Various instruments available for borehole survey are listed below.

Elinometer,
Mass compass,
Tropari drill hole surveying instrument,
Photographic angle recording devices,
Surwel gyroscope instrument and
Electronic and electrical surveying instruments.

Details on surveying are given in Chap. 5.

Underground Drilling: Surface drilling has been discussed earlier. The methods of drilling are much the same, but the scope of underground drilling is a little different. Directional, particularly horizontal and upward vertical and various inclined drill holes are more common in underground drilling. Besides, sub-surface drilling is done under many constraints, the important ones being the space restriction, lowering and hoisting of drilling equipment and the motive power required for driving the machinery. Therefore, the machinery and equipment necessary for underground drilling are different from those used for surface drilling. Generally Ex-core drills or electric drills are employed in underground operations.

Coring as well as non-coring drilling is done underground. Collection of core, sludge and cuttings is done as in surface drilling except in the case of vertical and inclined upward drilling where special equipment is necessary for collecting cuttings.

Underground drilling for finding new mica pegmatities is being done in Indian mica fields with extension rods in a normal jack hammer used for drilling blast holes. The cuttings so collected are examined for signs of mica mineralisation.

Exploratory Mining: In many cases of disseminated and vein-like deposits, surface exploration alone is not sufficient to get reliable data. Even in some massive deposits exposed on the surface, it becomes often necessary to get detailed data about the nature of the ore and the ore to waste contacts underground. Subsurface exploration is necessary in such cases. In order to learn more of a particular deposit, exploratory mining is resorted to in a detailed manner in the intensive exploration stage. Such mining may be either open-cast or underground. Open-cast methods are used for massive ore deposits like iron ore, bauxite, etc., and may consist of a few benches for studying the behaviour of the ore in the faces. Copper, lead, zinc, mica and such other minerals are explored mostly by underground or sub-surface methods.

As mentioned earlier, the common sub-surface exploration methods are aditing, drifting, cross-cutting, raising, winzing and underground drilling. All these methods cannot, however, be claimed as exclusively exploration methods. Drifts, cross-cuts, raises and winzes are put up for purposes of mining, blocking of ore and simple underground connections. However, whenever such openings show ore, observations, are possible. In fact, in the early development stage of an ore body these openings are made more often to expose ore.

Aditing: An adit is a horizontal or near horizontal opening driven into the ore body from some surface so that a cross-section of the ore body is available for study and sample collection. In large ore bodies, a number of adits may be required to study several cross-sections. If the ore body is amenable only to underground exploitation, the adit should be put at intervals of two to three times the height of the prospective levels.

Three types of adits are recognized, viz. adits driven along the strike of the ore body, adits driven along the dip of the ore body, and adits with blind shafts at the lowest horizon for very deep seated ore bodies.

Drifting: Drifting or driving is the process of making a sub-surface opening into the ore body along the strike. A drive may be along the footwall or hanging wall or both, or entirely through the ore body only depending upon the size and distribution of ore and the nature of the hanging and footwall contacts. Where a footwall or hanging wall contact is not clearly definable, the drives may be planned along that wall to define the ore body clearly. When the ore body is narrow but with well-defined contacts, a drive along the ore body may be more appropriate. The drive may start from the bottom of a shaft or other opening from the surface and may continue till the end of mineralization is in sight. Drives are generally planned in such a way that they can be made use of in exploitation at a future date. The number of such drives and their spacing will depend on a large number of variables and no guidelines are possible.

Cross-cutting: Cross-cutting is the process of driving an opening across an ore body exposed in an underground opening. A cross-cut may be made from a hanging wall drive to a footwall drive or vice versa. A number of cross-cuts might be necessary to study the ore in full width cross-section in a series of drives.

Raising and Winzing: Raises and winzes are vertical openings made to connect various levels. An opening made from a lower to a higher level is a raise and from a higher level to a lower level is a winze. All raises and winzes need not be in ore bodies. But some are put in the ore body purely for exploratory purposes.

3.18 Sampling

Sampling is done to ascertain the grade of a mineral and metal value that varies in proportion from one place to another. One single sample taken from one part of the ore body generally does not provide a representative picture of the grade of the entire ore body. A large number of well-spaced samples are required for ascertaining the average grade with an acceptable amount of accuracy. Normally, no amount of sampling will give a truly representative picture of the ore body. There is always some degree of error between the actual value and the value computed from the samples. The aim of sampling is only to reduce the error to the minimum possible level.

In addition to knowing the grade of the ore, sampling also reveals the pattern of mineralization within the ore body. A systematic mine sampling programme can demarcate the richer and leaner ore portions. Similarly, the limits of mineralization towards both the hanging and footwall contacts can also be precisely defined by careful sampling.

Sampling is also necessary to determine the processing and extractability characteristics of the ore. For this purpose, bulk representative/simulated samples representing the quality and type of material to be treated are collected.

3.18.1 Principles

A sample should be truly representative of the entire ore body. In order to attain this, it is necessary to choose proper places for sampling and it is always necessary to choose proper places for sampling and it is always necessary to show the sample sites and the width on the plan of the property. Any sample representing a very rich ore portion or a lean portion of the ore body loses its representative character. Theoretically, different samples collected from various parts of the ore body can be combined into a single composite sample to give the most representative picture of the whole ore body. This is never done because it is also necessary to know the average grade of the rich and lean portions of the ore body separately.

Samples should generally be taken so that at least all the exposed portions of the ore body are sampled. For this, samples may be spaced at regular intervals. The actual intervals cannot be determined arbitrarily, but have to be arrived at based on the experience gained in similar deposits in the past. It is not unusual for a gold-bearing quartz to be sample at every 50 cm in an underground drive. But an iron exposure of the type common in Indian need not be sampled at less than 50 m intervals. As a thumb rule, an assumption that any interval which minimizes the error is justified in sampling may not be very much out of place. It is a good practice to have a fixed minimum width or their multiples for each sample, depending upon the complexity of value distribution visualized.

Sampling is subject to certain limitations due to sampling errors. The errors may be of two types: (i) random and (ii) systematic. Of these, the random errors tend to cancel out each other whereas the systematic error accumulates to create gross errors which are easily seen because of their magnitude. The errors accumulate due to four factors.

(i) When check samples are taken from the same spot, there will be a natural divergence between the value of the principal sample and the control sample. This cannot be overcome,
(ii) Errors accumulated due to measurement errors, poor facilities and equipment and poor eye judgment of the sampler,
(iii) Errors due to mistakes of calculations, misprints and poor numbering and
(iv) Limitations of the assay technique itself.

Of these, the first two are random errors and the other two systematic. In addition, errors may crop up because of intentional or unintentional salting of the sample itself. All these errors have to be avoided to the extent possible to get a reliable estimate of the ore body. Check sampling and repeated sampling help in avoiding some of the mistakes whereas great care at every stage of sampling along can offset the other mistakes like salting.

3.18.2 Types of Sampling

In exploration, four types of sampling are of recognized value. They are chip sampling, grab sampling, channel sampling and bulk sampling. Occasionally, a geologist may have to resort to some special sampling techniques like car sampling, R.O.M. sampling, stack sampling, muck sampling, etc. 9. But these are of use only in specific situations. Sampling may be done for the determination of specific gravity, physical properties, petrological and mineralogical characteristics of the ore.

Chip Sampling: When values are regularly distributed as in an iron ore outcrop, chip sampling can be very useful. In chip sampling, first the outcrop or face to be sampled is cleaned properly and a regular, rectangular or square pattern is made by drawing lines and along and across the outcrop at fixed intervals. The, small pieces of ore are broken loose either from the centre of the grid or rectangle or at the intersection points of the lines. The ore pieces should have approximately the same shape, size and weight. After collecting a piece from each centre or intersection point of the grid, the rock pieces are mixed together to form the sample. In case of highly unpredictable values, the practice of shifting the grid by half the width or length of the grid is adopted to get another set of samples which may be mixed with the first set of chip samples and tested separately.

Grab Sampling: This is done from different blasts at the faces or small stacks of ore or dumps at random for information of a very general nature. Care should be taken to see that material of varying sizes is collected according to its proportion by weight in the blasted material or the stacks or dumps as the case may be. Several such grabs are mixed together to form a sample. In sampling by this method, the quantity of material to be collected depends on the size of the largest pieces present in the material to be sampled and the degree of heterogeneity of the material.

Channel Sampling: In this sampling, a channel is cut across the face of the exposed ore, and the resultant cuttings and chips are collected as a sample. The surface to be sampled is first cleaned to remove the dust, soluble particle, etc. A thin layer of the exposed ore may be removed to avoid cutting the weathered ore. Then a channel outline 5–10 cm in width and extending from the footwall to the hanging wall of the ore body is drawn by chalk or paint. Then the channel is cut by a moil and hammer to a depth which should be equal to the width. The resultant pieces are collected carefully on a clean sheet of canvas or any other convenient receptacle. The sides and the floor of the channel should be smooth and uniform so that overcutting (and overrepresentation) is effectively minimized 18. The channel may be divided into 1–2 m sections or their multiples in the case of massive and more homogeneous ore bodies or sections of 30–50 cm in the case of more heterogeneous distributions and may be separated as per the physical characteristics of the ore, say hard ore and laminated ore in the case of iron ores.

Bulk Sampling: Bulk sampling is done in two specific cases. One situation is when a pilot plant test is to be done on an ore. The other is when the constituents of the ore have to be determined accurately. Bulk samples may be made by collecting a portion from every blast continuously, or from shovels or cars in the case of mines. Bulk samples may be collected from a series of pits or a number of trenches, adits or underground drives in the case of prospects. For technological studies covering laboratory scale beneficiation tests, the bulk sample may be 100–250 kg in weight. In some complex ores, up to 100 kg may be necessary whereas for pilot plant tests 50 tonnes of material would be usually required.

Dump Sampling: Dump sampling can be done by systematically driving auger into the dumps and collecting the augured material. Benches may be prepared on the dumps and, from the benches, pits can be derived to collect samples. If a shovel is available, shovels can be deployed to take out representative bulk samples.

3.18.3 Criteria for the Selection of a Sampling Procedure for a Particular Mineral Type

The usual mineral sampling methods have been discussed above. Each method has its advantage and disadvantage. Therefore, some methods are very well suited to some type of deposits. The process of matching a deposit with the best sampling procedure suited for it requires certain criteria which are discussed below. Essentially, the criteria centre around the shape and type of mineralization of the deposit.

(i) When the ore body is thick and the values of mineralization are uniformly distributed, sampling can be done by chip or grab sampling.
(ii) When the ore body is of medium size and mineralization is uniform, a combined chip and channel sampling will give the best results.
(iii) Where the ore body is too thin but occurs in benches or layers, sampling of various layers can be done by chip sampling.
(iv) When a deposit is of very large dimension, it becomes necessary to collect a large number of samples. In such cases, a large number of chip samples would give reliable results. Here, the quicker and not necessarily the most accurate method should be preferred.
(v) With the ore is banded, channel sampling would give the best results and
(vi) Very hard ore, particularly massive types of iron ore, would require to be sampled by blasthole cuttings.

3.18.4 Spacing of Sample Channels

There are no rigid rules regarding spacing of sample channels in any type of deposit. However, spacing can be controlled by mathematical analysis, which is discussed in Chap. 6. Notably, the formula using the standard error of statistical mean and the confidence interval is eminently suited for predetermining the channel spacing and the number of samples required for specific precision limit.

3.18.5 Collection of Samples

The collection of samples is a job requiring skill and experience. All chips, blocks and powder coming from a groove should be gathered irrespective of the size of concentration of mineral value. No extraneous material should get mixed up with

Table 3.25 Suggested guideline for assay register

Name of the Mine:										
Sample No.	**From**	**To**	**Assay width**	**True width**	**Assay value X**	**Geological information**				
					In X metals	Lithology	Type of mineralisation	Strike	Dip	Remarks
						Weighted average grade from X1 to X0 width				

the sample. The sample should be collected in a clean canvas bag. After the completion of a groove, the collected chips/blocks, etc., should be put in bag and properly labelled by marking on the outside of the bag and also putting a reference tag inside the bag. A field book of sample records containing serially numbered—sheets with a guideline for descriptions, and arrangement to retain the counterfoil of description in the book, and a similarly numbered label part going into the sample bag will be desirable. A proper register showing the location of the sample co-ordinates, channel logs, sample weight, time taken for sampling, etc., should be maintained by the sampler. The register should show the serial numbers of the samples. It is always preferable to complete the register as soon as a sample is collected. Table 3.25 gives a guideline for the maintenance of a register.

We have seen that economic geology encompasses a variety of activities. One important aspect of these disciplines is the identification and evaluation of mineral deposits. Exploration geologist uses the tools of geology, geophysics and geochemistry, and other fields to identify mineral deposit. We than study the deposits to determine if they exhibit the physical and chemical characteristic of similar deposits that have been developed into mines, and in case of new types of deposit, to draw inferences on the basis of new knowledge. The other parts are, however, economic and equally important to start a mine. To evaluate a mineral deposits economic potential by comparing the expected revenues from mine production with the associated costs of further exploration, development and production. Other aspects that encompasses are socioeconomic in nature.

Further Reading

Arogyaswami RNP (1988) Course in mining geology, 3rd edn. Oxford IBH, p 695

Brooks RR (1972) Geobotany and biogeochemistry in mineral exploration. Harper Row Publishers Inc., New York

Carranza EJM (2008) Geochemical anomaly and mineral prospectivity mapping in GIS. In: Handbook of exploration and environmental geochemistry, vol 11. Elsevier Publication, Amsterdam, p 351

Colin ED (2007) Biochemistry in mineral exploration. In: M. Hale (ed) Handbook of exploration and environmental geochemistry. Elsevier Publication, Amsterdam, p 462

Dentith M, Midge ST (2014) Geophysics for mineral exploration geoscientist. Cambridge University Press, Cambridge, p 454

Dobrin MB (1960) Introduction to geophysical prospecting. McGraw Hill, New York

Edwards R, Atkenson K (1986) Ore deposit geology. Chapman and Hall, London, p 466

Gocht H, Zaruop RG, Eggernt PG (1988) International mineral economics. Springer, Berlin, p 271

Govett GJS (1983) Handbook of exploration geochemistry. In: Rock geochemistry in mineral exploration, vol I. Elsevier, Amsterdam, p 461

Guilbert JM, Park CF (1985) The geology of ore deposits. WH Freeman, New York, p 985

Harris DP (1984) Mineral resources appraisal. Oxford University Press, Oxford, p 445

Kearey P, Vine FJ (1992) Global tectonics. Blackwell Scientific, p 302

Kearey P, Brooks M, Hill I (2002) An introduction to geophysical exploration. Blackwell Science, p 262

Levinson AA (1974) Introduction to exploration geochemistry. Allied Publication (with 2nd ed., 1980 supplement), p 924
Marjoribanks R (2012) Geological methods in mineral exploration and mining. Springer, Berlin
Morse JG (2013) Nuclear methods and mineral exploration and products, Elsevier, Amsterdam, p 292
National Research Council of the USA (1981) Mineral resources: genetic understanding for practical application studies in geophysics, p 119
Rose AW, Hawkes HE, Webb JS (1979) Geochemistry in mineral exploration. Academic Press, p 657
Rossi M, Deutsch CV (2014) Mineral resource estimation. Springer, Berlin, p 327

Chapter 4
Remote Sensing in Mineral Exploration

4.1 Concept

Remote sensing was first adopted as a technique for obtaining information of distance objects without being in physical contact to the object (Fisher 1975). In practical terms remote sensing collects electromagnetic or acoustic signals. Earliest aerial were taken in 1858 from a balloon. However, since 1930 aerial photography using aircraft has been used extensively for resources survey. Satellite photography of the earth has been available to geologists since the early 1960s. Several hundred high oblique satellite photographs were acquired with a 70 mm hand-held camera during one of the Mercury missions in 1961. First formal geologic photography experiment (Gemini Mission) was carried out in 1965 (Lowman 1969). The multispectral terrain photography experiment on Apollo 9, which used four Hasselbled cameras and four different film–filter combinations, acquired 90 sets of photographs on 70 mm film between 3 and 13 March 1969. The success of the Apollo 9 experiment set the stage for both the Landsat programme and the Skylab Project. Operational usage of remote sensing were adopted and continue at present with the launch of remote sensing satellites like Landsat series (USA), SPOT series (France), Skylab(USA), IRS series (India), ERS series (Europe), MOS series (Japan), JERS-l (Japan) and ASTER (Japan and USA). Several high resolution satellites were also launched that could be used in geological high resolution mapping projects.

Remote sensing has gained importance due to its applications in various mineral exploration scenarios. The ore bodies may be associated with specific rock types, controlled by geological structures, and associated with different hydrothermal types. Some of geologically economic deposits such as laterite, bauxite show different morphological features. Remote sensing can help the exploration geologist to distinguished different rock types, map the hydrothermally altered rocks, investigate the morphological features and analyze the structural features.

G.S. Roonwal, *Mineral Exploration: Practical Application*,
Springer Geology, DOI 10.1007/978-981-10-5604-8_4

Today we utilize the remote sensing as a useful method in mineral exploration, since remote sensing helps in locating ore bodies that show up some surficial features such as geomorphology, hydrothermal alteration, structural features, lithology and geobotany.

4.2 Remote Sensing System

The link between the components of the remote sensing system is the electromagnetic (EM) energy. Electromagnetic energy disperse as wave energy with the velocity of light ($c = 3 \times 10^{10}$ cm/see). Here, we know that the source of EM radiation maybe originating from natural sources such as sun's reflected light or heat emitted by earth, or man-made, like microwaves. The physical or compositional properties of surface materials determine the amount and characteristics of the EM radiation. The sun is the principal source of EM radiation. The EM radiations incident on the earth's material can get absorbed or reflected. The reflectance can be specular, diffused and scattered. Remote sensing data can be classified based on the wavelength. The visible region spans between 0.4 and 0.7 μm. Near-infrared region is 0.7–1.0 μm (VNIR). Short-wave infrared (SWIR) begins from one micrometre and ends with three micrometres. The thermal infrared (TIR) considered to be 3–20 μm. The microwave region is between 1 cm and 1 m. Hence, we can classify remote sensing systems into optical and radar remote sensing.

The energy received by the sensor is then converted to analogue or digital form either onboard the spacecraft/airborne sensor or on the ground. The raw data received by the ground station usually needs to be preprocessed in order to make them usable for the end users (e.g. atmosphere and geometric corrections) (Lillesand and Keiffer 2000).

4.2.1 Remote Sensing Sensors

Remote sensing sensors acquire data in different wavelengths that is called spectral band. Each spectral band registers electromagnetic radiation in specific band region of electromagnetic spectrum. There are several remote sensing satellites that are imaging the Earth's surface. Although digital images of Landsat, ASTER, IRS and SPOT are well known amongst the geologists, but ASTER and Landsat images are more popular due to their spectral capabilities for geological mapping.

4.2.1.1 Landsat and ASTER Systems

Landsat data have mostly been used semi-arid or desert setting. It helps in locating for example iron oxides (gossan/oxidation) or even hydrous minerals such as

Table 4.1 Characteristics of ETM+ and Landsat-8 images

	Landsat 7	Landsat 8
Spectral bands	7	11
VNIR resolution (m)	30	30
SWIR resolution (m)	30	30
TIR resolution (m)	120	100
PAN resolution (m)	15	15
Spectral bands	μm	μm
Band-1	0.45–0.52	0.433–0.453
Band-2	0.52–0.60	0.450–0.515
Band-3	0.63–0.69	0.525–0.600
Band-4	0.76–0.90	0.630–0.680
Band-5	1.55–1.75	0.845–0.855
Band-6	10.4–12.5	1.56–1.66
Band-7	2.08–2.35	2.1–2.3
Band-8	0.5–0.9	0.5–0.68
Band-9	–	1.36–1.39
Band-10	–	10.30–11.30
Band-11	–	11.50–12.50

gypsum or even hydrothermal alteration. Table 4.1 shows the characteristics of different bands of ETM+ and Lansat 8 images.

The ASTER is an imaging instrument on the Terra platform. The satellite is a cooperative effort between NASA and Japan's Ministry of International Trade and Industry. ASTER obtains information on surface emissivity, temperature, reflectance and elevation. ASTER obtains data in 14 spectral channels from the visible through thermal infrared regions of the EM. It consists of three separate instrument subsystems (Abrams et al. 2002). Individual bandwidths and subsystems characteristics are summarized in Table 4.2. Theoretically, the SWIR bands of ASTER

Table 4.2 ASTER characteristics (Abrams et al. 2002)

Band	VNIR spectral resolution (mm)	Band	SWIR Spectral resolution (mm)	Band	TIR spectral resolution (mm)
1(nadir)	0.52–0.60	4	1.600–1.700	10	8.125–8.475
2(nadir)	0.63–0.69	5	2.145–2.185	11	8.475–8.825
3(nadir)	0.76–0.86	6	2.185–2.225	12	8.925–9.275
3(back ward)	0.76–0.86	7	2.235–2.285	13	10.25–10.95
		8	2.295–2.365	14	10.95–11.65
		9	2.360–2.430		
Technology detector	Pushbroom Si		Pushbroom PtSi:Si		Whiskbroom Hg: Cd:Te
Spatial resolution (m)	15 × 15		30 × 30		90 × 90
Swath width	60 km		60 km		60 km
Quantization	8 bits		8 bits		8 bit

have more capability than the Landsat for recognition of areas with hydrothermal alteration.

4.2.1.2 Indian Remote Sensing Satellites (IRS Series)

In India Remote Sensing is conducted by in Indian Space Research Organization's (ISRO)—Indian Remote Sensing satellites (IRS) series Earth observation satellites. Today there are also OceanSat, CartoSat, ResourceSat and others in the series. Table 4.3 shows specifications of these satellites.

Hyperspectral Imaging
The sensors that acquire several hundreds of spectral bands over a single area are called hperspectral imagers. These images are capable of mapping mineralogical composition present at the surface of the Earth. Thus, the airborne visible-infrared imaging spectrometer (AVIRIS) and HyMap are the airborne hyperspectral sensors which are capable of acquiring as many as 200 images over a single area.

The EO-1 (Earth observation) Hyperion is the only available space borne hyperspectral data available till today (Kruse et al. 2003). The imageries for this sensor is available free at www.USGS.gov. You may find lots of information related to characteristics of the sensors onboard EO-1 from this website.

Characteristics of Satellite Images
Each spectral band consists of columns and rows of pixels. Every pixel has an address in an image that can be shown as BVijk. BV stands for brightness value for each pixel, *i* and *j* represent column and row numbers, respectively, and k represents the spectral band number. For example in Fig. 4.2, the highlighted pixel in band-1 can be addressed as $52_{1822,\ 696,\ 1}$.

Figure 4.1 shows a subset of six spectral bands of ETM+ data over an area that contains vegetation cover (with black pixels in bands 1 and 3) and a hydrothermal alteration (with bright pixels in band 5). An enlarged portion of this image that is shown with a red square is shown in Fig. 4.2. The corresponding BVs or digital numbers (DNs) are also shown. Every surface feature may absorb electromagnetic energy in one wavelength and reflects in another. Therefore, in a grey scale image the features are in shades of grey. For example, the vegetation cover that is seen with dark pixels in bands 1, 2, 3 and 7; appears in higher DNs in band 4. This is due to the fact that green vegetation cover has absorptions in bands 1, 3 and 7 and reflection in band 4 of ETM+ images. The pixels covering the hydrothermally altered areas have higher BVs in band 5 and lower BVs in band 7. This phenomenon can be used in later sections for enhancing vegetation cover and hydrothermally altered rocks.

Table 4.3 Major specifications of IRS series of satellites (Navalgund et al. 2007)

Satellites (year)	Sensor	Spectral bands (μm)	Spatial res. (m)	Swath (km)	Radiometric res. (bits)	Repeat cycle (days)
IRS-IA/1B (1988, 1991)	LISS-I	0.45–0.52 (B) 0.52–0.59 (G) 0.62–0.68 (R) 0.77–0.86 (NIR)	72.5	148	7	22
	LISS-II	Same as LISS-I	36.25	74	7	22
IRS-P2 (1994)	LISS-II	Same as LISS-I	36.25	74	7	24
IRS-1C/1D	LISS-II	0.52–0.59 (G), 0.62–0.68 (R) 0.77–0.86 (NIR) 1.55–1.70 (SWIR)	23.5 70.5 (SWIR)	141 148	7	24
	WiFS	0.62–0.69 (R) 0.77–0.86 (NIR)	188	810	7	24(5)
	PAN	0.50–0.75	5.8	70	6	24(5)
IRS-P3 (1996)	MOS-A MOS-B MOS-C WiFS	0.755–0.768 (4 bands) 0.408–1.010 (13 bands) 1.6 (1 band) 0.62–0.68 (R) 0.77–0.86 (NIR) 1.55–1.70 (SWIR)	1570 × 1400 520 × 520 520 × 640 188	195 200 192 810	16 16 16 7	24 24 24 5
IRS-P4 (1999)	OCM MSMR	0.402–0.885 (8 bands) 6.6, 10.65, 18, 21 GHz (V & H)	360 × 236 150, 75, 50 and 50 km, respectively	1420 1360	12 –	2 2
IRS-P6 (2003)	LISS-IV	0.52–0.59 (G) 0.62–0.68 (R) 0.77–0.86 (NIR)				

(continued)

Table 4.3 (continued)

Satellites (year)	Sensor	Spectral bands (μm)	Spatial res. (m)	Swath (km)	Radiometric res. (bits)	Repeat cycle (days)
	LISS-III	0.52–0.59 (G), 0.62–0.68 (R) 0.77–0.86 (NIR) 1.55–1.70 (SWIR)	5.8 23.5	70 141	10(7) 7	24(5) 24
	AWiFS	0.52–0.59 (G) 0.62–0.68) (R) 0.77–0.86 (NIR) 1.55–1.70 (SWIR)	56	737	10	24(5)
IRS-P5 (Cartosat-I) 2005	Pan (Fore (+26°) & Aft (−5°))	0.50–0.85	2.5	30	10	5
Cartosat-2 (2007)	PAN	0.50–0.85	0.8	9.6	10	5

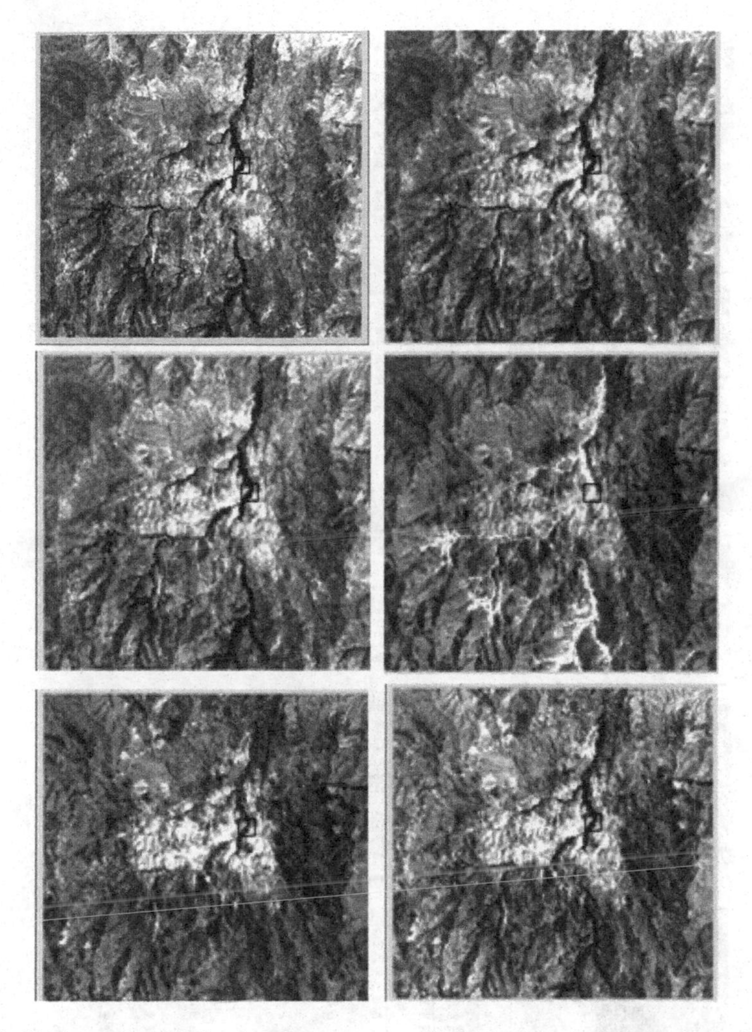

Fig. 4.1 Six images of ETM+ that show an area with hydrothermal alteration and vegetation cover

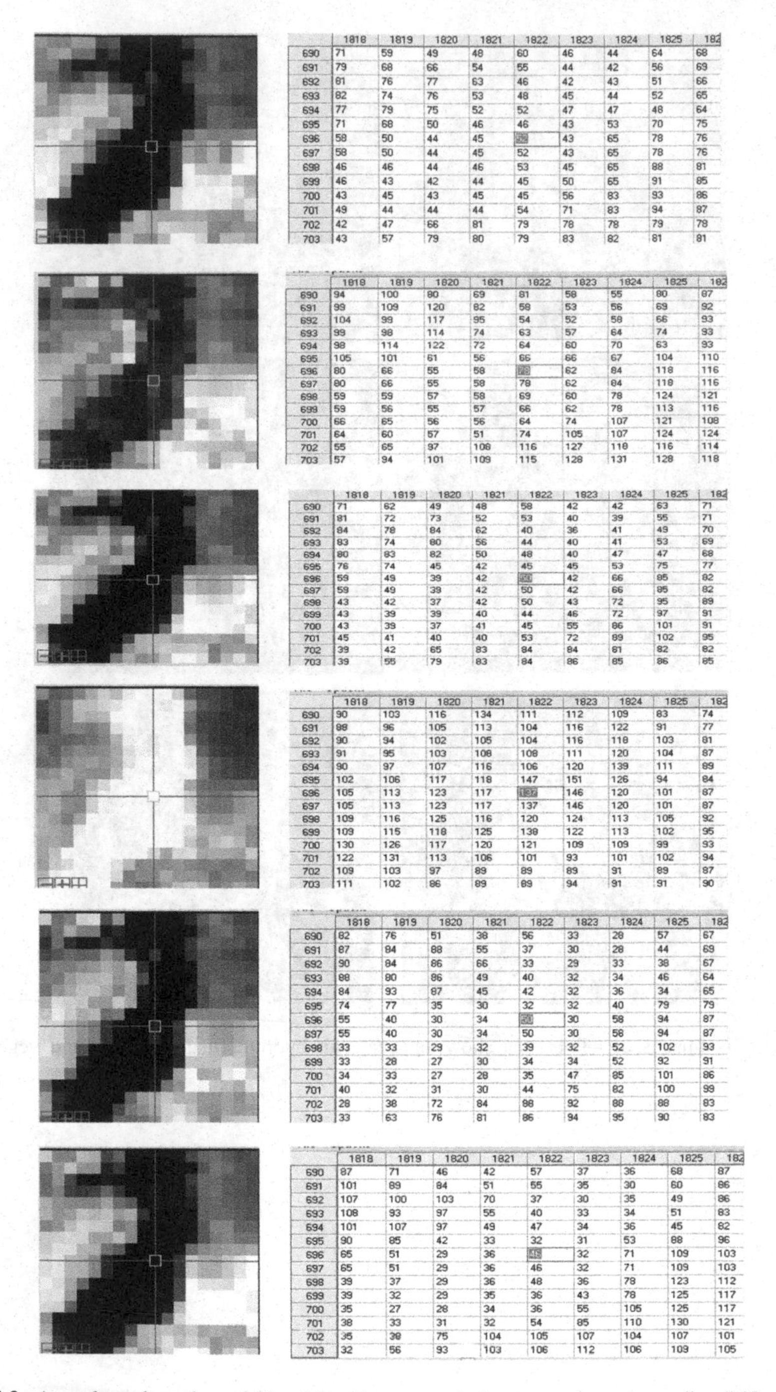

	1818	1819	1820	1821	1822	1823	1824	1825	182
690	71	59	49	48	60	46	44	64	68
691	79	68	66	54	55	44	42	56	69
692	81	76	77	63	46	42	43	51	66
693	82	74	76	53	48	45	44	52	65
694	77	79	75	52	52	47	47	48	64
695	71	68	50	46	46	43	53	70	75
696	58	50	44	45	[illegible]	43	65	78	76
697	58	50	44	45	52	43	65	78	76
698	46	46	44	46	53	45	65	88	81
699	46	43	42	44	45	50	65	91	85
700	43	45	43	45	45	56	83	93	86
701	49	44	44	44	54	71	83	94	87
702	42	47	66	81	79	78	78	79	78
703	43	57	79	80	79	83	82	81	81

	1818	1819	1820	1821	1822	1823	1824	1825	182
690	94	100	80	69	81	58	55	80	87
691	99	109	120	82	58	53	56	69	92
692	104	99	117	95	54	52	58	66	93
693	99	98	114	74	63	57	64	74	93
694	98	114	122	72	64	60	70	63	93
695	105	101	61	56	66	66	67	104	110
696	80	66	55	58	[illegible]	62	84	118	116
697	80	66	55	58	78	62	84	118	116
698	59	59	57	58	69	60	78	124	121
699	59	56	55	57	66	62	78	113	116
700	66	65	56	56	64	74	107	121	108
701	64	60	57	51	74	105	107	124	124
702	55	65	97	108	116	127	118	116	114
703	57	94	101	109	115	128	131	128	118

	1818	1819	1820	1821	1822	1823	1824	1825	182
690	71	62	49	48	58	42	42	63	71
691	81	72	73	52	53	40	39	55	71
692	84	78	84	62	40	36	41	49	70
693	83	74	80	56	44	40	41	53	69
694	80	83	82	50	48	40	47	47	68
695	76	74	45	42	45	45	53	75	77
696	59	49	39	42	[illegible]	42	66	85	82
697	59	49	39	42	50	42	66	85	82
698	43	42	37	42	50	43	72	95	89
699	43	39	39	40	44	46	72	97	91
700	43	39	37	41	45	55	86	101	91
701	45	41	40	40	53	72	89	102	95
702	39	42	65	83	84	84	81	82	82
703	39	55	79	83	84	86	85	86	85

	1818	1819	1820	1821	1822	1823	1824	1825	182
690	90	103	116	134	111	112	109	83	74
691	88	96	105	113	104	116	122	91	77
692	90	94	102	105	104	116	118	103	81
693	91	95	103	108	108	111	120	104	87
694	90	97	107	116	106	120	139	111	89
695	102	106	117	118	147	151	126	94	84
696	105	113	123	117	[illegible]	146	120	101	87
697	105	113	123	117	137	146	120	101	87
698	109	116	125	116	120	124	113	105	92
699	109	115	118	125	138	122	113	102	95
700	130	126	117	120	121	109	109	99	93
701	122	131	113	106	101	93	101	102	94
702	109	103	97	89	89	89	91	89	87
703	111	102	86	89	89	94	91	91	90

	1818	1819	1820	1821	1822	1823	1824	1825	182
690	82	76	51	38	56	33	28	57	67
691	87	84	88	55	37	30	28	44	69
692	90	84	86	66	33	29	33	38	67
693	88	80	86	49	40	32	34	46	64
694	84	93	87	45	42	32	36	34	65
695	74	77	35	30	32	32	40	79	79
696	55	40	30	34	[illegible]	30	58	94	87
697	55	40	30	34	50	30	58	94	87
698	33	33	29	32	39	32	52	102	93
699	33	28	27	30	34	34	52	92	91
700	34	33	27	28	35	47	85	101	86
701	40	32	31	30	44	75	82	100	99
702	28	38	72	84	88	92	88	88	83
703	33	63	76	81	86	94	95	90	83

	1818	1819	1820	1821	1822	1823	1824	1825	182
690	87	71	46	42	57	37	36	68	87
691	101	89	84	51	55	35	30	60	86
692	107	100	103	70	37	30	35	49	86
693	108	93	97	55	40	33	34	51	83
694	101	107	97	49	47	34	36	45	82
695	90	85	42	33	32	31	53	88	96
696	65	51	29	36	[illegible]	32	71	109	103
697	65	51	29	36	46	32	71	109	103
698	39	37	29	36	48	36	78	123	112
699	39	32	29	35	36	43	78	125	117
700	35	27	28	34	36	55	105	125	117
701	38	33	31	32	54	85	110	130	121
702	35	38	75	104	105	107	104	107	101
703	32	56	93	103	106	112	106	109	105

Fig. 4.2 An enlarged portion of Fig. 4.1 with *grey* scale images and corresponding DNs

4.2.2 EM Spectrum

Electromagnetic spectrum is a continuum consisting of the ordered arrangement of radiation according to the wavelength, frequency or energy. It extends from highly energetic cosmic rays photons through gamma rays, X-rays, ultraviolet, visible, infrared, microwave and radiowaves. The wavelengths that are greatest interest in remote sensing are visible and near-infrared radiation.

4.2.3 Spectral Characteristics of Hydrothermal Alteration Minerals

A group of minerals could be used as index minerals that occur in the altered rocks associated with various mineral deposits such as porphyry copper deposits. The spectral properties of minerals can be used for their identification, based on their reflectance behaviour. The spectral reflectance characteristics of rocks and minerals in the visible near-infrared (VNIR) through the short-wave infrared (SWIR) wavelength regions (0.4–2.5 μm) are the result of different physical and chemical properties. However, the spectral feature that is typically displayed by the well-defined bands is caused by absorptions due to both electronic and vibration processes in the individual mineral constituents (Hunt and Salisbury 1970; Hunt and Ashley 1979). Here, we are concerned with the process that leaves to locating the mineral deposits. Between 0.35 and 1.3 μm, electronic transitions in the iron-bearing minerals (hematite, goethite and jarosite) cause characteristic features, in the form of minima, to occur near 0.43, 0.65, 0.85 and 0.93 μm. They are common components in many ore minerals and indicating iron oxide-rich caps,

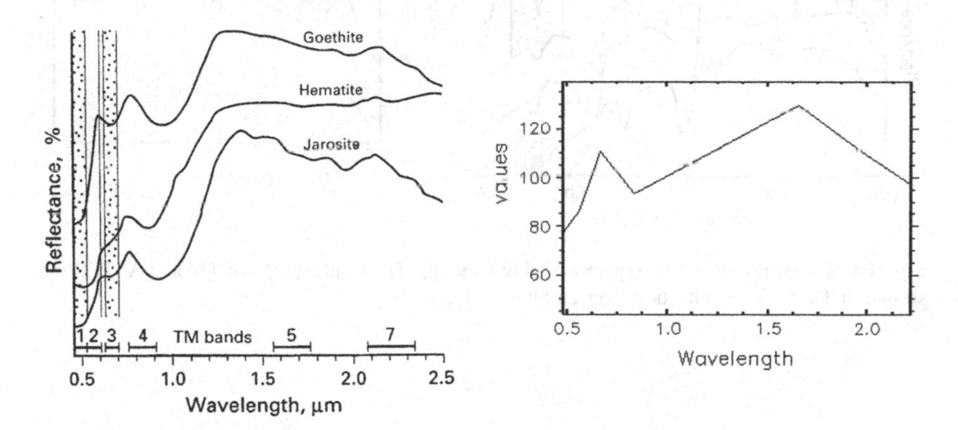

Fig. 4.3 **a** Spectral pattern of oxidized zone—gossan for jarosite, hematite and goethite and **b** Spectral profile on an area rich in iron oxide. Note the peak in band 3 and an absorption feature in band-1 of ETM+ data

known as gossan. Figure 4.3 depicts the spectra of three iron oxide bearing samples that show absorption in blue band (band 1) due to electronic process (charge-transfer) and reflections in red and near-infrared (NIR) portions (bands 3 and 4). This explains why we see oxidize zone in form of gossan which contain hematite in red and brown.

The oxidation process and products generated comprise within it hydrous (OH) molecules or even carbonate associated minerals of Al, Fe, Mg as aluminium ore, iron ore or MgOH. They constitute clays, sulphates and water-bearing minerals (alunite, kaolinite, illite, montmorillonite, Pyrophyllite, mica, diaspore, jarosite, chlorite and carbonates) that are common in the hydrothermally altered rocks (Hunt and Ashley 1979). Figure 4.4 depicts the spectrum. These mineral have an absorption feature within band 7 and a reflection in band 5 of ETM+, respectively. Here, the bands ETM+ and ASTER sensors can distinguished between clay minerals and alunite, while, ETM+ images is not able to do so, due to the fact that band 7 has more width and covers all absorption features.

Since ASTER and Landsat data are widely used by the geologist around the world more than any other satellite systems, here the spectral properties of rocks and minerals are explained based on the spectral bands of these data. The ASTER, VNIR + SWIR pattern of mineral generated due to hydrous nature of mineral can be seen clearly. These mineral comprise micas (sericite) illite in rock such as phyllite. The rock basically consists of clays—kaolinite (Fig. 4.5).

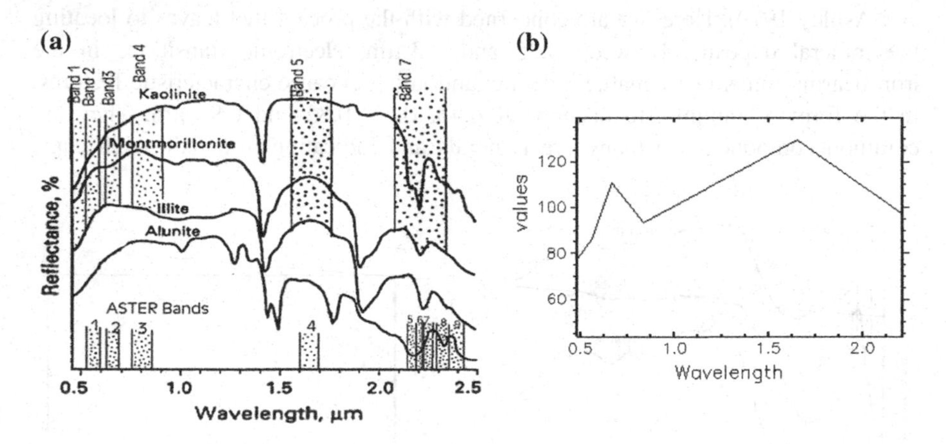

Fig. 4.4 **a** Showing reflectance spectra of clay minerals. The bands widths of TM and ASTER are shown. **b** The relative reflection over an altered area

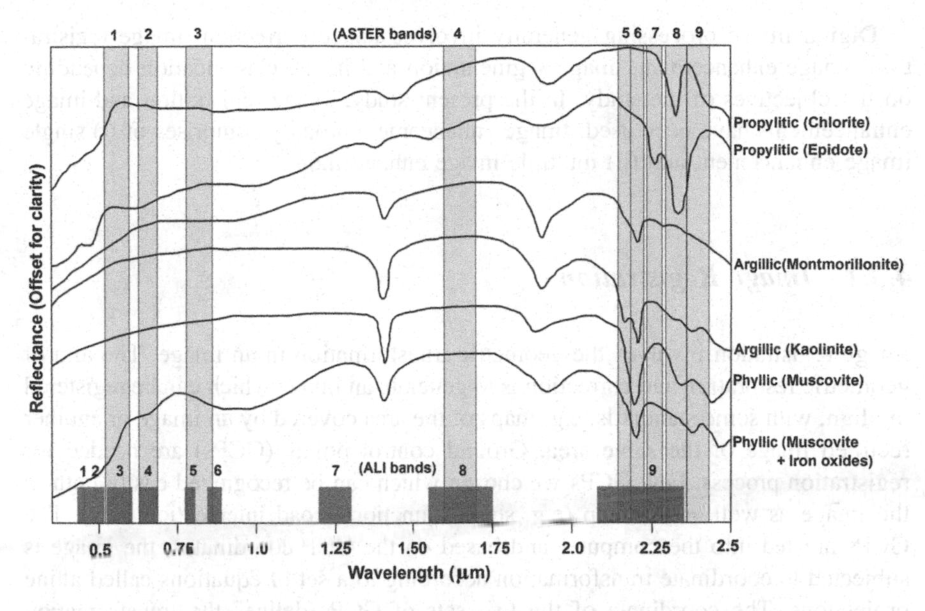

Fig. 4.5 Showing a Laboratory Reflectance Spectra of important hydrothermal alteration minerals. ASTER and ALI bands are shown in figure (Honarmand et al. 2011)

4.3 Discrete or Digital Image Processing of Satellite/Airborne Images

Gillespie (1980), Hord (1982, 1986), Swain and Davis (1978), Lillesand and Kiefer (1987, 1994), Niblack (1986), Mather (2001), Jensen (1996) and Gupta (2003) have given reviews in digital image processing of remote sensing data. A digital image comprises a number of individual picture elements called Pixels, each one having an intensity value (Digital number) and an address in two dimensional image space, i.e. rows and columns The digital number (DN) may represent the reflectance of EM radiation (albedo, emissivity, temperature) or some geophysical, geochemical or topographical data. The DN value is dependent of the intensity range of the image which usually stretches from minimum (0) to maximum (255 in 8 bit format). The image data are usually stored in computer compatible tapes (CCT), floppy disks, data cartridges, digital audio tapes (DAT) and CD-ROM medium.

We understand that the digital image processing is carried out by computer processing of digital data generated. In general the purpose of digital image processing is to enhance or improve the image quality to extract information from it. Digital image processing offers precision and flexibility over optical or electrical methods. Voluminous data becomes a challenging task to analyze within a limited time. Use and application has helped greatly in data processing and time saving. Many useful digital image processing operations are available on personal computers and desktop workstations (Niblack 1986).

Digital image processing generally involves image correction, image registration, image enhancement, image segmentation and image classification depending on the objectives of the study. In the present study, image registration and image enhancements have been used. Image enhancement broadly comprises of (i) single image enhancement and (ii) multiple image enhancement.

4.3.1 Image Registration

Image registration involves the geometric transformation in an image. The aim of geometric restoration and correction is to generate an image which can be registered or align, with some standards, e.g. maps of the area covered by an image or another rectified image of the same area. Ground control points (GCPs) are needed for registration process. Few GCPs we chosen which can be recognized easily both in the image as well as the map (e.g. stream junctions, road intersection, etc.). The GCPs are fed into the computer and based on the GCP coordinates, the image is subjected to coordinate transformation according to a set of equations called affine projections. The coordinate of the two sets of GCPs defines the transformation parameters. Typically a set of two equations (affine projections) is used to link the two coordinate systems:

$$X' = a0 + a1x + a2y + a3xy$$
$$Y' = b0 + b1x + b2y + b3xy$$

where X' and Y' are the coordinates in the new system, and x and y are the coordinates of the same points in the map. $a1$, $a2$, $a3$, $b0$, $b1$, $b2$ and $b3$ are the unknown constants which can be computed using four control points. To use the above equations to geometrically transform an image four stages are required. First, a geometrically correct geographical grid is defined in terms of latitude, longitude or northing and easting. Second, the computer proceeds through each cell in this geographical grid and at each cell the computer transforms the latitude/longitude or northing/easting values into values of x and y which becomes the new address of an image pixel. Third, the computer visits this address in the image and transfers the appropriate DN by interpolation. Forth, this process is repeated until the geographical grid is full at which point the image has been geometrically correlated (Bernstein 1983).

4.3.2 Image Enhancement

Image enhancement involves the processes applied on the images in order to improve the image quality so as to make them more interpretable. By applying a particular enhancement technique some features may become better discernible at the cost of other features, e.g. directional filtering enhances the linear features at a particular directions though other features in the image is suppressed. A particular band ratio may enhance vegetation cover and another ratio may enhance the iron oxide or argillic alteration. There are several methods of image enhancement, such as (1) contrast stretching, (2) band rationing, (3) principal component analysis, (4) addition and subtraction, (5) RGB coding, (6) spectral angle mapper and (7) image filtering.

4.3.3 Contrast Stretching

Here, monitor screen of processor, or the light source in a digital film writer is from 0(zero-black) to 255(maximum intensity), but in general, pixels in an image often occupy a small portion of the possible range of grey-level values, resulting in low-contrast display on which some features might be indistinguishable. Image processing software can change any DN in an image to any of 255 intensity levels. The stretching is done by spreading the DNs equally over the 0–255 range. The minimum DN is set to 0 and the maximum to 255 with other DNs falling between these two extreme values. Guides to distinguished or identified even minor variation is observed. Different methods of contrast stretching such as linear contrast stretching, multiple linear stretching, logarithmic, power or functional stretching, Gaussian stretching, histogram equalization stretching and density slicing are available in the modern image processing softwares. In all cases the images are stretched.

4.3.4 Band Rationing

Band rationing or spectral rationing is an extremely useful procedure for enhancing features in the multispectral images. In this technique, the DN values of one band is divided by the corresponding DN values of another band, pixel by pixel, and the resultant data is rescaled to fill the dynamic range of the display device by contrast stretching operation. Ratio images are useful because they have the effect of suppressing the detail in a scene which is caused by topographic effects (i.e. variable effects of illumination conditions) while enhancing colour boundaries. This property has made ratio pictures quite useful in geological applications because they exaggerate subtle colour differences in a scene and many geological problems require distinction between rock types that may appear quite similar. Ratio images are interpreted because they can be directly related to the spectral properties of materials. More information can be obtained by using those ratios that maximize the differences in the spectral slopes of materials in the scene. Rationing technique has been used to enhance argillic verses non-argillic, rock versus vegetation, iron oxide

Table 4.4 ASTER band ratio for enhancing mineral features (based on van der Meer et al 2012)

Mineral feature	ASTER bank combinations(2)
Ferric iron	2/1
Ferrous iron	5/3 and 1/2
Ferric oxide	4/3
Gossan	4/2
Carbonate/Chlorite/Epidote	(7 + 9)/8
Epidote/Chlorite/Amphibole	(6 + 9)/(7 + 8)
Amphibole	(6 + 9)/8 and 6/8
Dolomite	(6 + 8)/7
Carbonate	13/14
Sericite/Muscovite/Illite/Smectite	(5 + 7)/6
Alunite/Kaolinite/Pyrophyllite	(4 + 6)/5
Phengite	5/6
Kaolinite	7/5
Silica	11/10, 11/12, 13/10
SiO_2	13/12, 12/13
Siliceous rocks	(11 × 11)/(10 × 12)

versus non iron oxide (Drury and Hunt 1989). A disadvantage of rationing is that it suppresses differences in albedo. The rocks such as basalt and marls will appear the same in the ratio image, though they have a lower and a higher albedo, respectively. Table 4.4 presents few band ratios for enhancing mineral features.

It is possible to use, apart from the simple rationing, the output images resulted from the other image enhancement techniques (e.g. PC images, ISH images, added and subtracted images, etc.) as either denominator or numerator. This technique may provide in some cases, useful information for geological studies.

4.3.5 Addition and Subtraction

Subtraction and addition of satellite images are simple and useful methods of image enhancement when the multispectral images are highly correlated. Addition of spectral images generates an image with much larger dynamic range than original images. Therefore, higher contrast image is the result. The image which is produced by differences of two images is characterized by lower contrast. This technique particularly enhances the areas which are less correlated in the original images and therefore it is possible to derive change-detection image from multispectral data. Image subtraction sometimes can give the same result as image ratio but with much simpler operation (Navai and Mehdizadeh-Tehrani 1994).

4.3.6 Principal Component Analysis (PCA)

The principal component analysis is a multivariate technique that selects uncorrelated linear combinations (eigenvector loadings) of variables in such a way that each successively extracted linear combination, or PC, has a smaller variance. Remove redundancy from the multispectral data is the main aim of PC analysis. Principal component analysis is extensively used for mapping of hydrothermal alteration in metallogenic provinces (Abrams et al. 1983; Loughlin 1991; Tangestani and Moore 2001, 2002).

A feature oriented principal component selection is known as Crosta technique. It allows identification of the principal components (through the analysis of the eigenvector values) that contain spectral information about specific minerals, as well as the contribution of each of the original bands to the components in relation to the spectral response of the materials of interest. This technique can be applied on four and six selected bands of Thematic Mapper (TM) data. The technique indicates whether the materials are represented as bright or dark pixels in the principal components according to the magnitude and sign of the eigenvector loadings.

4.3.7 Red-Green-Blue (RGB) Coding

The colour images are the result of the three additive primary colours (red, green and blue) in the RGB colour coordinate system. However, it is often difficult to choose proper band combination and produce an optimum FCC especially in the cases where several images are involved. Optimum index factor (OIF) may solve this problem up to some extent (Chaves et al. 1982). The technique uses the computation of total variance and correlation within and between bands. The combination with higher OIF is chosen for making false colour composite. Similar statistical method suggested by Hunt et al. (1986) and according to this method the most informative three spectral bands are the least well correlated ones.

4.3.8 Spectral Angle Mapper (SAM)

The Spectral Angle Mapper (SAM) allows mapping of the spectral similarity of image spectra to reference spectra. The reference spectra can be selected from either laboratory or field spectra or extracted from the image. Spectral analysis assumes a spectral angle which represents the data has been reduced to apparent reflectance, with all dark current and path radiance biases removed. SAM compares the angle between the reference spectrum and each pixel vector in n-dimensional space, and smaller angles represent closer matches to the reference (Kruse et al. 1993). The angle can be calculated using following equation:

$$\propto = \cos^{-1}\left(\frac{\sum_{i=1}^{nb} i_i r_i}{\left(\sum_{i=1}^{nb} i_i^2\right)^{1/2}\left(\sum_{i=1}^{nb} r_i^2 r_i^2\right)^{1/2}}\right) \tag{4.1}$$

i_i and i_r are the image and reference spectra, respectively, and *nb* is the number of bands. Lower angle indicates more correlation between the image spectra and reference spectra.

The SAM algorithm has been used on hyperspectral and multispectral data for hydrothermal alteration mapping (Ranjbar and Honarmand 2007; Di Tommaso and Rubinstein 2007; Tangestani et al. 2008; Shahriari et al. 2013).

4.3.9 Image Filtering

Spatial filters emphasize or deemphasize image data of various spatial frequencies. Spatial frequency refers to the degree of changes in pixel values from one pixel to another. In a high frequency image, the tonal changes are abrupt (e.g. changes across lithological boarder). In contrast, a low frequency image has gradual tonal changes (e.g. tonal changes within a lithology or water body). Low pass filters are designed to emphasize low frequency features and deemphasize the high frequency components of an image (local detail). High pass filters emphasize the detailed high frequency components of an image and deemphasize the more general low frequency information. There are several low frequency filters such as mean, median, mode, Gaussian, etc. High pass filters include Laplacian, Sobel and Roberts. The high pass filters are of two types as follows:

1. The filters that enhance the high frequency features in all directions.
2. The filters that enhance the high frequency features in a specific direction.

For example, if enhancement of faults in a particular direction is desired, a directional filter can be applied. In cases where the lithological boundaries that may not have specific directions, Laplacian filter may be applied. Enhancement of drainage pattern in an image also can be done using Laplacian filter.

In order to digitally filter an image, a kernel is used. Figure 4.6 shows a Laplacian (A) and a high pass directional filters' kernels (B). The latter kernel can enhance the features in east–west direction. This kernel is moving over an image and a new value is calculated for the central pixel every time the kernel stops over nine pixels.

As mentioned earlier in the preceding chapters, many of ore deposits (such as porphyry and vein types) are associated with faults or lineaments. To enhance these features, the satellite images can be filtered with high pass filters. After this step, the

(a)

-1	-1	-1
-1	8	-1
-1	-1	-1

(b)

0	-2	0
0	4	0
0	-2	0

Fig. 4.6 Kernels for **a** Laplacian filter and **b** directional filter

lines showing the geological features should be drawn manually with a help of Geographic Information System (GIS) software.

Another aspect is the Photolineament Factor (PF) is applied for analyzing lineaments distribution and has application in mineral exploration, as suggested by Hardcastle (1995). Generally, in the system a grid with cell size (e.g. 2 × 2 km) is superimposed over the lineament map and each parameter is calculated from the respective cells. The photolineament factor value is calculated using the following equation:

$$PF = a + b + c \tag{4.2}$$

where 'a' is number of lineament intersections in each cell/average of the area, 'b' is number of lineaments in each cell/average of the area and 'c' is number of major lineament directions in each cell/average of the area. According to our study (Ranjbar and Roonwal 2002) the PF values are thus controlled. Our study further showed that in known occurrence of mineralization zones even in highly altered condition, this method is very useful.

4.4 Application of Remote Sensing in Mineral Exploration

As mentioned earlier, mineralization may be associated with a particular lithology, geological structure, morphology, hydrothermal alteration, etc. remote sensing may help mapping or enhancing these geological features especially in the arid/semi-arid parts of the world. Here, few geological features are investigated using remote sensing.

4.4.1 Geological Structures

It is an established fact that some of ore bodies, petroleum/gas reservoirs are controlled by geological structures. Here, several types of structures are shown that depict the ability of remote sensing for mapping them. Figure 4.7 shows a plunging syncline in the eastern part of Iran that hosts coal beds. The rocks in the syncline are composed of sandstone, shale, limestone and marl of Jurassic age. The coal beds are seen in the shale and sandstone sequence. The southern limb of the fold is normal but the northern limb is inverted as shown in Fig. 4.7. The fold is plunging to the southeast direction.

The host lithology is a sequence of shale and sandstone that is seen in bluish colour in Fig. 4.8 that goes round the fold. This image helps the geologists to locate the coal beds in the coal–sandstone sequence in other parts of the fold. We know that folded structures are proper places for exploration of hydrocarbon. Some of these structures are outcropping at the surface that can be mapped by remote sensing. If these structures are covered by alluvium or other lithologies, they can be mapped by geophysical methods such as gravity and seismic. Here remote sensing can help the geophysists to lay down proper geophysical lines for data collection (Ranjbar 2011).

Figure 4.8 shows many double plunging anticlines in south of Iran. The outcropping lithology is limestone that belongs to the Asmari and Jahrom Formations which act as reservoirs while being enclosed with impermeable strata.

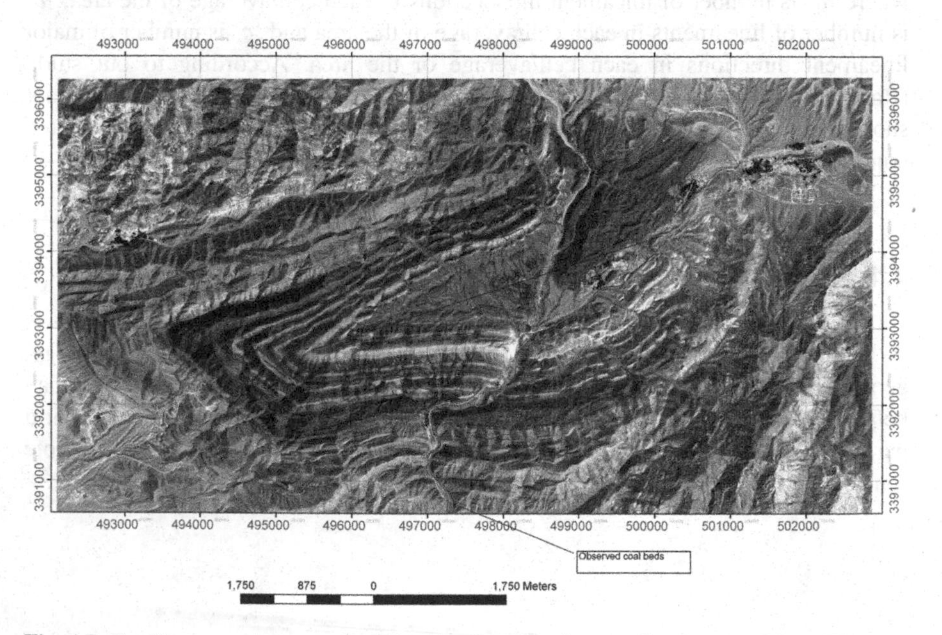

Fig. 4.7 Satellite image in true colour that shows a plunging syncline in east of Iran. The map coordinate is N, UTM, Zone 40, WGS84

Fig. 4.8 Landsat image in false colour that shows doubly plunging anticlines in south of Iran (color figure online)

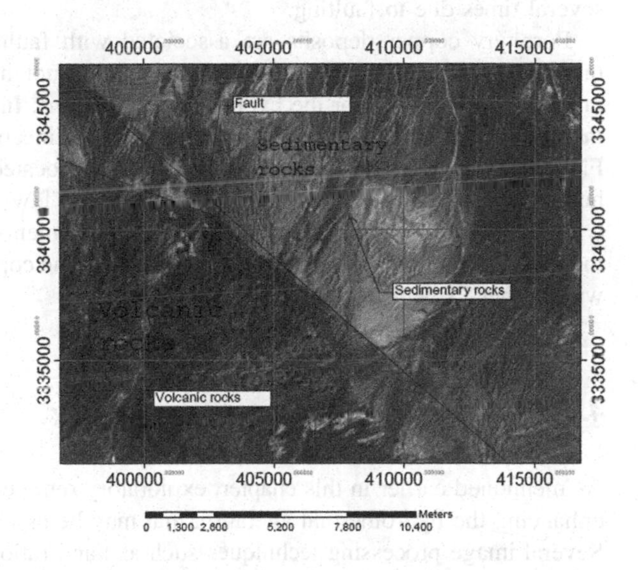

Fig. 4.9 Faults that separated volcanic from the sedimentary rocks. The map coordinate is N, UTM, Zone 40, WGS84

Here they are outcropping at the surface and do not contain hydrocarbon. Hydrocarbon may be present within the underlying strata. Faults are important structure in mineral exploration scenarios. They provide pathways for hydrothermal solutions, increase permeability, and act as barrier and traps for hydrocarbons.

Although the lithological changes are abrupt on both sides of a fault, it is very easy to detect it (Fig. 4.9). In cases where the fault has occurred in one lithology,

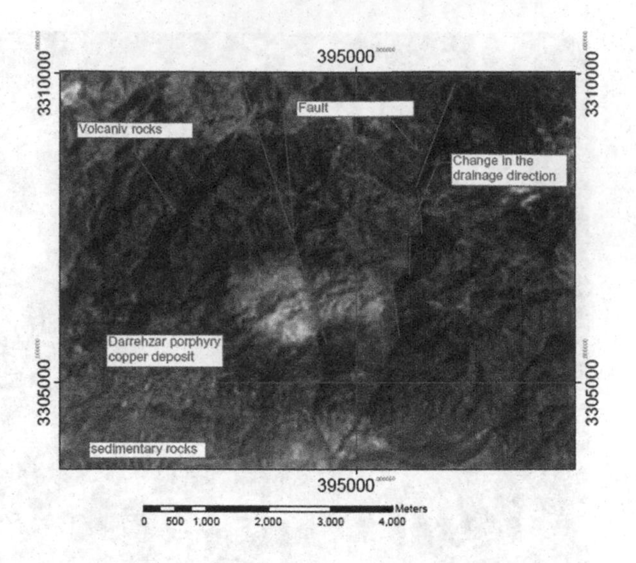

Fig. 4.10 A false colour image of landsat. Faults that cut across the Darrehzar intrusive body which hosts a porphyry copper deposit. The map coordinate is N, UTM, Zone 40, WGS84 (color figure online)

the changes in drainage pattern are used for its recognition. The faults are cutting across the Darrehzar intrusive which is a granodiorite that hosts a porphyry copper deposit in southeast of Iran (Fig. 4.10). On the eastern part, the drainage is dissected several times due to faulting.

Porphyry copper deposits are associated with faulting systems. The faulting increases permeability in the rock units that do not have primary permeability, provide the pathways for the hydrothermal solutions. In order to graphically show the areas with higher lineament density, photolineaments maps are prepared. Figure 4.11 depicts the Landsat image of an area located in the southeast Iran that hosts Sar Cheshmeh porphyry copper deposit and few smaller porphyry and vein type mineralization. This image is filtered, the lineaments were extracted and finally the PF value contours were drawn using Eq. 4.2. The copper deposits are associated with higher PF values.

4.4.2 *Hydrothermal Alteration Mapping*

As mentioned earlier in this chapter, exploratory remote sensing is mostly used for enhancing the hydrothermal alteration that may be associated with mineralization. Several image processing techniques such as band rationing, principal component analysis and spectral angle mapper can be used for mapping hydrothermal alteration types.

Porphyry copper deposits received large attentions for exploration by the remote sensing techniques all over the world. The associated zones of hypogene hydrothermal alteration and weathering are spatially large enough to be detected and mapped by using multispectral remote sensing data. Most of the known

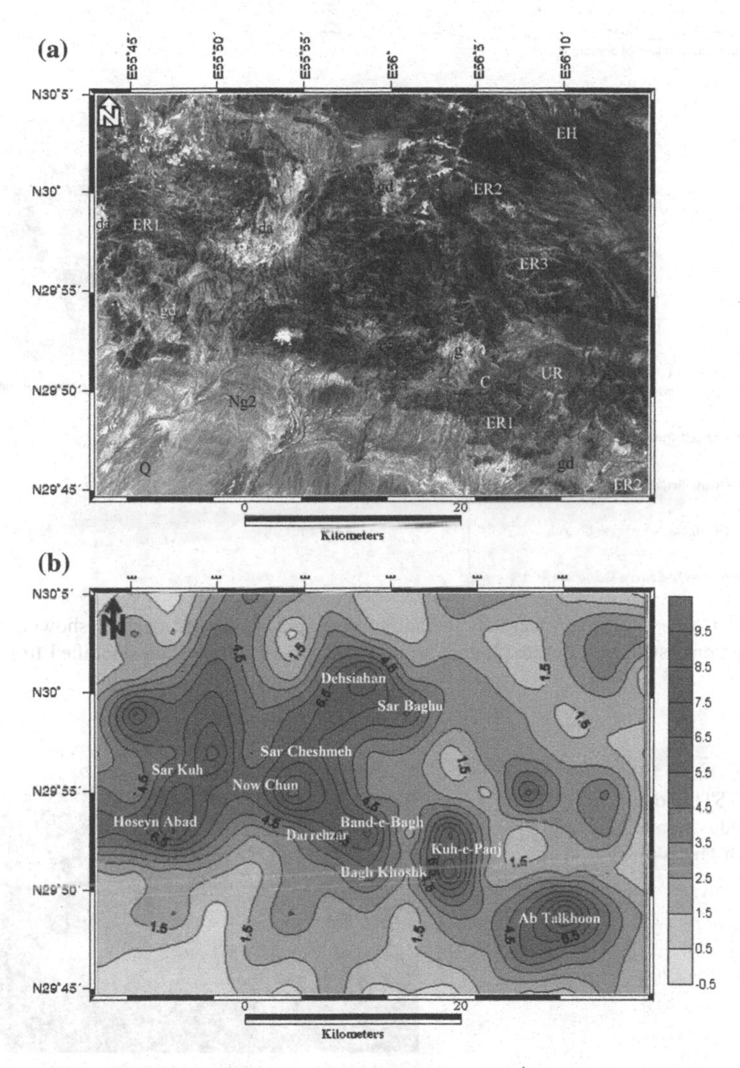

Fig. 4.11 **a** False colour image that shows lithological variation in the Sarcheshmeh area. EH, ER1, ER2, *ER3* volcanic rocks of Eocene, *gd* granodiorite, *dc* dacite, *Ng2* Neogene sedimentary rocks, *Q* Quaternary sedimens. **b** Photolineament factor value of the above image (color figure online)

Porphyry copper deposits are characterized by a well-developed zonal pattern of mineralization and wall rock alteration that can be defined by assemblages of hydrothermal alteration minerals. The most intense alteration occurs in the core of the porphyry body and diminishes radially outward in a series of concentric zones of alteration minerals (Fig. 4.12).

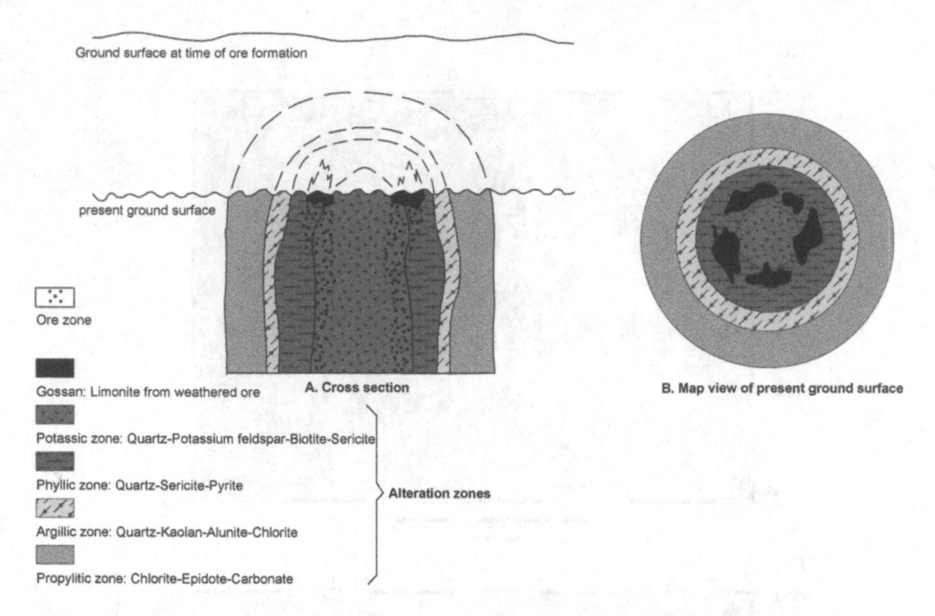

Fig. 4.12 Concept of hydrothermal alteration of porphyry copper deposit showing various rock-type zones such as, potasic, phyllic, argillic and propylitic alteration (modified from Sabins 1999)

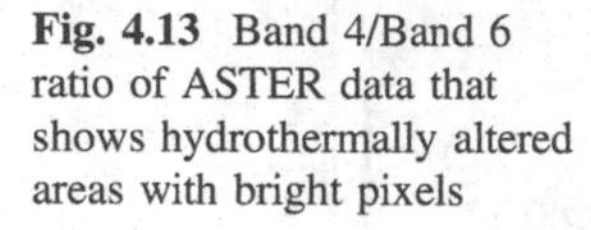

Fig. 4.13 Band 4/Band 6 ratio of ASTER data that shows hydrothermally altered areas with bright pixels

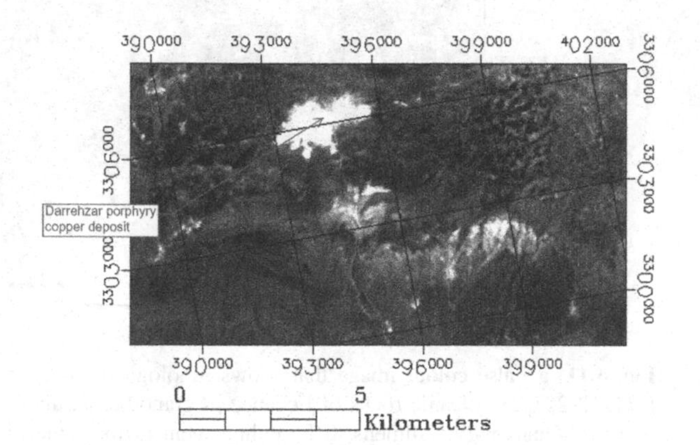

4.4.2.1 Band Ratio

It has been observed that hydroxide molecules of minerals form large halos because of hydrous nature. Examples of this are clearly seen in the minerals comprising sheet silicates since they contain Hydroxyl-bearing minerals form the most widespread products of hydrothermal alteration. An abundance of clays and sheet silicates, which contain Al–OH– and Mg–OH-bearing minerals and hydroxides in alteration zones, is characterized by absorption bands in the 2.1–2.4 μm due to

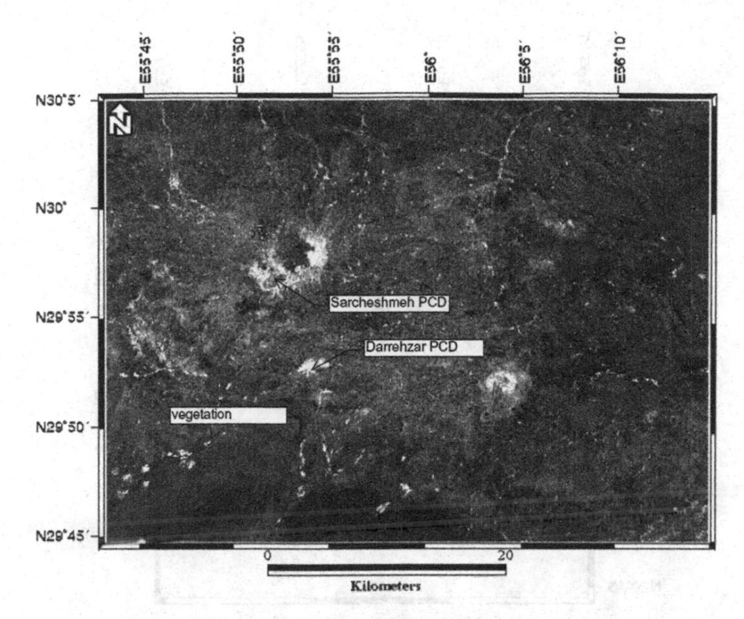

Fig. 4.14 Band 5/Band 6 of ETM+ data that depicts hydrothermal alteration with bright pixels

molecular vibrational processes (Fig. 4.4). Here, due to clays and sheet silicates higher reflectance in band 4 of ASTER data is observed. Here, common features of phyllosilicates have been recorded and utilized for mineral exploration (e.g. Galvao et al. 2005; Hubbard and Crowley 2005; Mars and Rowan 2006; Rowan et al. 2006).

Band ratio techniques though simple are important criteria in distinguishing and identifying remote sensing image analysis. It guides to spectral differences between bands as well as to help reduce influence of topography (Rowan and Mars 2003; Zhang et al. 2007). Distinguishing spectral bands produce an image that provides relative band intensities. The resulted image enhances the spectral differences between the bands. Areas with hydrothermal alteration are usually enhanced as bright pixels in images of band 4/band 9, band 4/band 5 and band 4/band 6 for clay and sheet silicates and 9/8 ratio for chlorite and epidote in ASTER data. Figure 4.13 shows band4/band6 for Darrehzar porphyry copper deposit. Figure 4.14 shows band 5/band 7 ratio images of Landsat ETM+ data. In both images the hydrothermally altered areas are enhanced with bright pixels. The area of this image covers the area shown in Fig. 4.11.

4.4.2.2 Principal Component Analysis

In Sect. 4.3.6 we have already discussed about principal component analysis, additional aspects of PCA are narrated here.

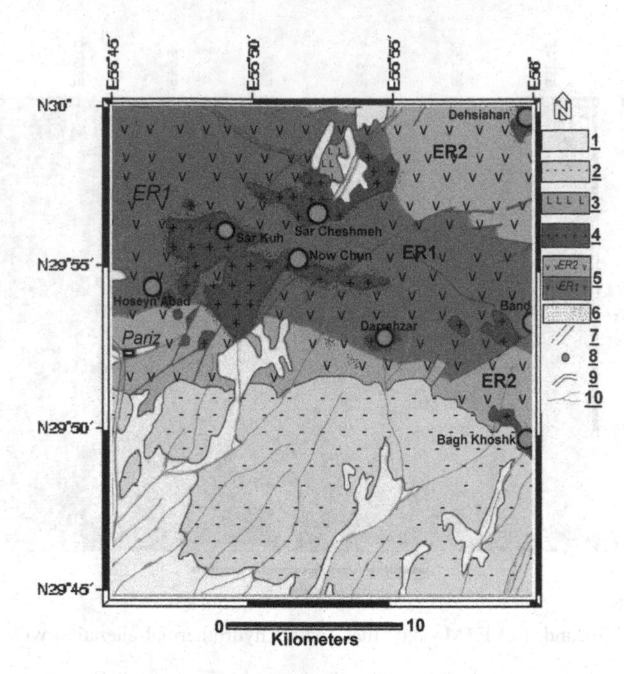

Fig. 4.15 Geology of porphyry copper deposit of Sar Cheshmeh, Kerman, Iran. Index shows: *1* Quaternary alluvium, *2* Quaternary gravel fan, *3* Quaternary calcareous terraces, *4* Neogene sediments, mostly arenites with pebbles and boulders of volcanic and intrusive rocks. Dacites and dacitic pyroclastics, *5* Granodiorite, quartz diorite, diorite porphyries and monzonite, dikes of Oligocene-Miocene age, *6* Eocene volcanic-sedimentary complex, trachyandesites, trachybasalts, basaltic andesites, pyroclastics, etc. *7* Fault, *8* Working mine and copper deposit, *9* hydrothermal alteration (after, Dimitrijevic et al. 1971)

Table 4.5 Eigenvector loadings for six bands of ETM+ data

	PC1	PC2	PC3	PC4	PC5	PC6
Band 1	0.26	−0.49	0.20	0.02	0.62	−0.50
Band 2	0.34	−0.39	0.18	−0.21	0.14	0.79
Band 3	0.49	−0.39	0.02	0.09	−0.72	−0.26
Band 4	0.34	0.010	−0.91	0.11	0.20	−0.37
Band 5	0.53	0.54	0.13	−0.62	0.05	0.05
Band 6	0.42	0.40	0.29	0.74	0.13	−0.16
% of variance	88	8.5	1.9	0.08	0.05	0.01

Standard PCA is applied on ETM+ data of Sar Cheshmeh area. The intrusive bodies are composed of diorite, quartz diorite and granodiorite of Oligocene–Miocene age that intrude Eocene Volcanic-Sedimentary complex comprised mainly of volcano-clastics, andesite, trachy-andesite and sedimentary rocks. The hydrothermally altered rocks are highly fractured, and supergene alteration has

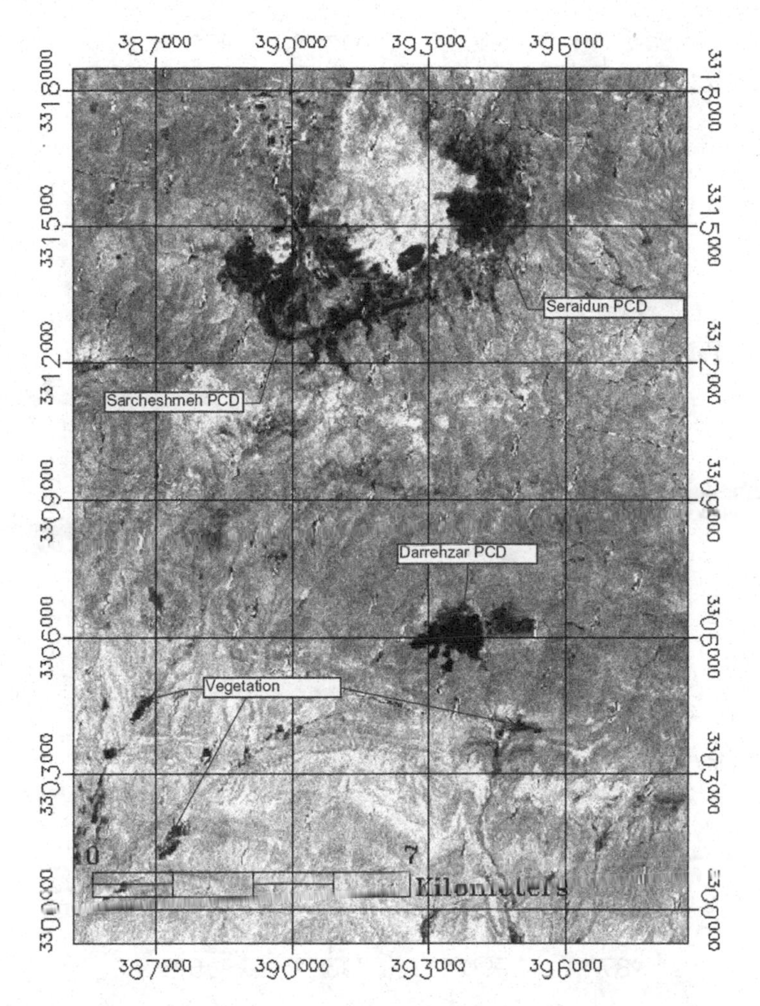

Fig. 4.16 PC4 image that shows altered areas with *black pixels*. The map coordinate is N, UTM, Zone 40, WGS84

produced extensive limonite and leaching of sulphide, giving a characteristic reddish or yellowish colour to the altered rocks. A weathered zone is developed a few metres to 80 metres below the surface (Fig. 4.15).

The eigenvector loadings and the eigenvalues are described in Table 4.5, using six ETM+ spectral bands. The first PC contains 88% of the variance of the six bands and gives information mainly on albedo and topography as all the bands have positive loadings. PC3 enhances vegetation cover in dark pixels, as the loading for band 9 is high and negative. PC4 enhances the hydroxyl minerals as both bands 5 and 7 have high loadings with opposite signs. Negative value for band 5 loading causes the altered parts to appear as dark pixels in Fig. 4.16. The vegetation cover also appears as dark pixels because of their water content that cause absorption in

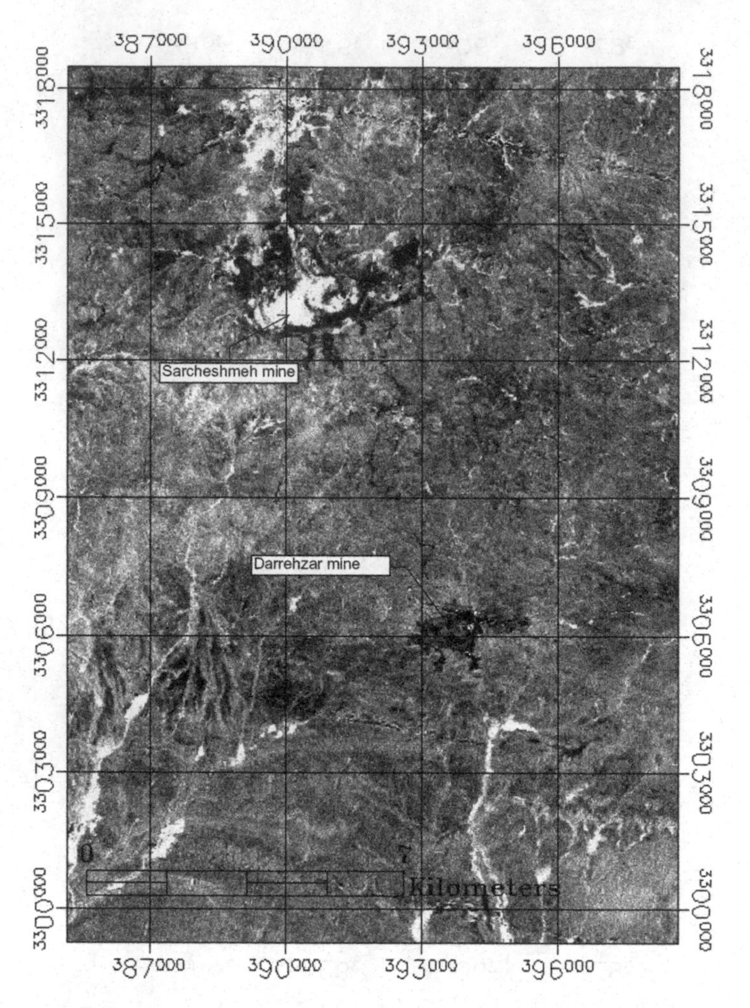

Fig. 4.17 PC4 image that shows iron oxide bearing areas with *black pixels*. The map coordinate is N, UTM, Zone 40, WGS84

band 7 of ETM+. PC5 enhances iron oxide minerals. Iron oxide minerals have higher reflectance in band 3 and absorption in band 1 of ETM+. Here both the bands have high loadings but with opposite signs. As the reflectance band has negative signs, the areas with iron oxide minerals appear in dark pixels (Fig. 4.17). Gossan or iron oxide cap rock is developed over the mineral deposits that have undergone the process of oxidation. Elements such as copper, molybdenum, Pb, Zn, etc. are leached out and iron oxide minerals such as hematite, goethite and jarosite are formed at the surface. This can be an indication of buried deposit. If you compare Figs. 4.16 and 4.17, it is noticed that the areas with hydrothermal alteration also contain iron oxide minerals. The Sar Cheshmeh mine pit has alteration, but iron oxide minerals are extracted out and the mining activity is now in the supergene and hypogene parts. That is the reason that iron oxide is not seen over the mine pit.

4.4.2.3 Spectral Angle Mapper

Since our focus is on mineral exploration to map hydrothermal alteration haloes around porphyry copper mineralization, the spectra of selected alteration minerals from USGS spectral library were used as the end members and SWIR bands of ASTER data were used in the analysis. Muscovite (sericite) and illite is representative of phyllitic zone; kaolinite and montmorillonite are representative of argillic zone; and chlorite and epidote are representative of propylitic zone. Let's apply SAM method on Darrehzar area. The ASTER SWIR bands should be preprocessed before applying SAM. Internal average relative reflectance was applied on SWIR images. The preprocessed SWIR bands classified using SAM method (Fig. 4.18).

SAM classification method gave a good result that was very close to the reality for identification of alteration types in the area. When, we compare Figs. 4.13 and 4.18, it is clear that SAM classification has provided far better result than ratio image.

Govil (2015), used SAM method to map hydrothermally altered minerals around Askot basement mineralization of Kumoan Himalaya, India, using EO-1 hyperspectral data. Askot basemetal mineralization occurred in the Askot crystallines of the Kumaon Himalaya, India. In the Kumaon Himalaya at Askot copper deposit, the country rocks are crystallines which are underlain by younger formation of Inner Sedimentary Belt (ISB) of the region. Structurally, this is confined by Main Central Thrust in the north, and Almora Thrust in the South. As mentioned the crystallines rock comprise varieties of gneisses such as augen gneiss, granite gneiss,

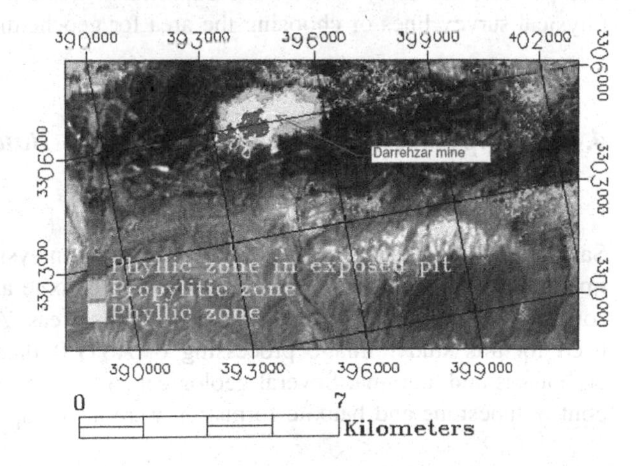

Fig. 4.18 Classified ASTER SWIR images using SAM. The map coordinate is N, UTM, Zone 40, WGS84

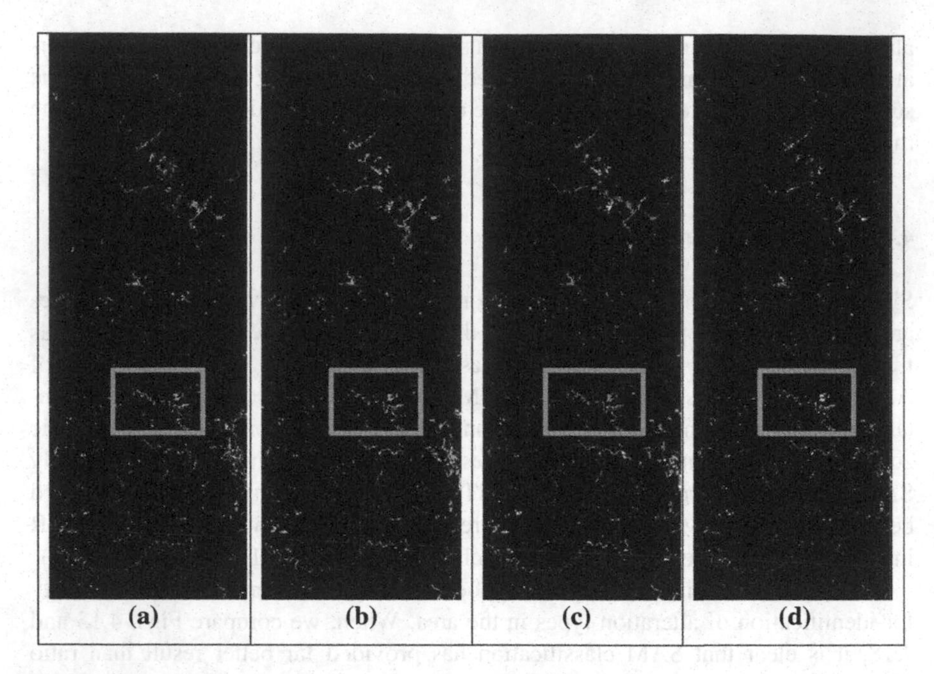

Fig. 4.19 Mineral map of the Askot basement mineralization. **a** Chlorite, **b** Goethite, **c** Illite, **d** Muscovite (Govil 2015)

garnetiferous-biotite gneiss, biotite-muscovite gneiss. In addition, calcsilicates and quartzites are also seen. Dykes of aplites and pegmatites are frequent (Govil 2015).

The Hyperion data was preprocessed and later on analyzed using SAM method (Govil 2015). Figure 4.19 shows the resulted images that depict the distribution of alteration minerals in the area.

These alteration maps providing the initial exploratory data that acts as basis for other exploratory techniques such as geophysics, geochemistry and drilling. They can provide necessary data to the exploration geologist for laying out the geophysical survey lines or choosing the area for geochemical sampling.

4.4.3 Application of Remote Sensing in Bauxite and Carbonate Exploration

Sanjeevi (2008) studied the potential of spectral analysis of multispectral satellite image data for targeting of mineral content in bauxite and limestone rich areas in southern India around Ariyalur and Kolli Hills areas. ASTER images have been used for this study. Image processing of ASTER data delineated areas rich in carbonates and alumina. Several geological and geomorphological parameters that control limestone and bauxite formation were also mapped using ASTER images.

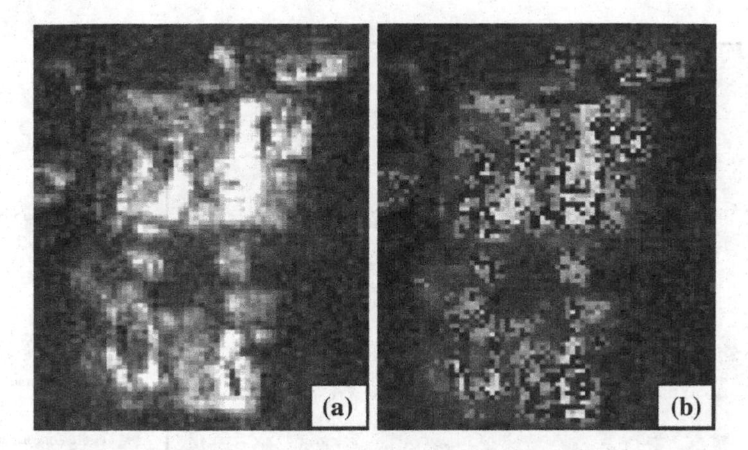

Fig. 4.20 **a** Carbonate fraction image and **b** density sliced fraction image; *red* <50%; *green* 50–60%; *blue* 60–70%; *yellow* 70–80%; *cyan* 80–90%; *white* >90% (Sanjeevi 2008) (color figure online)

Fig. 4.21 Density sliced alumina fraction image draped over digital elevation model, showing location with 70–100% alumina (shown in *red*) (Sanjeevi 2008)

Figure 4.20 shows an image that shows $CaCO_3$ abundances which is derived from ASTER data. Figure 4.21 shows abundance map of bauxite that is derived from ASTER data.

4.4.4 Application of Remote Sensing in Exploration of Placer Deposits

In the southern state of Tamil Nadu in the peninsular India along the coast line occurs excellent deposit of heavy mineral beach placers. These deposits have been investigated by remote sensing (Fig. 4.22). Beside mineral exploration, remote

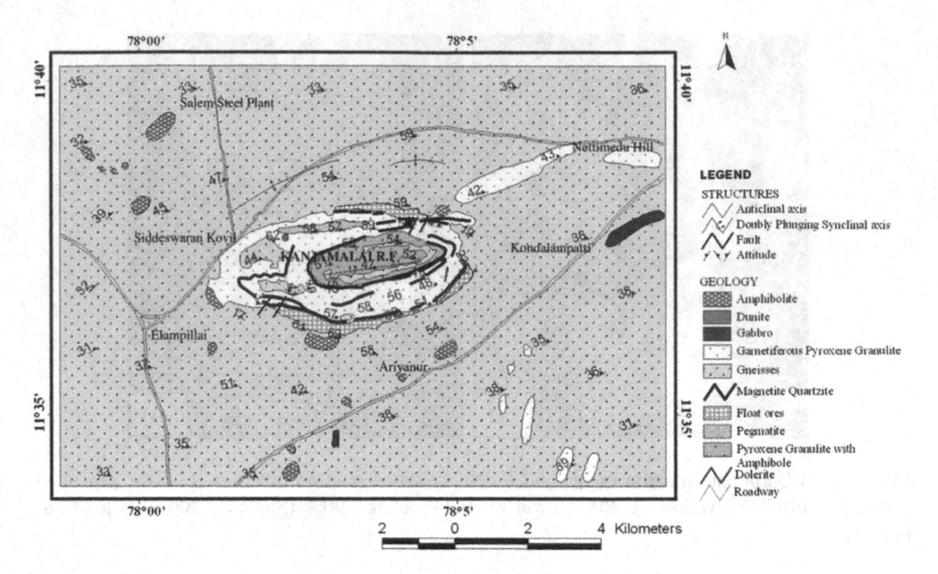

Fig. 4.22 Lithological map of Salem region (Rajendran et al. 2011)

sensing has also helped to identify environmental parameters for sustainable mining of these deposits. It has also guided to control coastal erosion.

4.4.5 Application of Remote Sensing for Iron Ore Exploration

In the high-grade granulite country around Salem, Rajendran et al. (2011) described a technique for discriminating iron ores (magnetite quartzite deposits) and associated lithology in high-grade granulite region of Salem, Southern Peninsular India using visible, near-infrared and short-wave infrared reflectance data of Remote Sensing—ASTER Image spectra has shown lithology very clearly. They comprise magnetite quartzite of garnetiferous pyroxene granulite, hornblende biotite gneiss, amphibolite, dunite and pegmatite have absorption features around spectral bands 1, 3, 5 and 7.

Deposits of iron ores here are banded iron formations rich in iron and iron silicates of meta-sedimentary rocks. A full geological succession is visible in Kanjamalai region of Salem (Fig. 4.23).

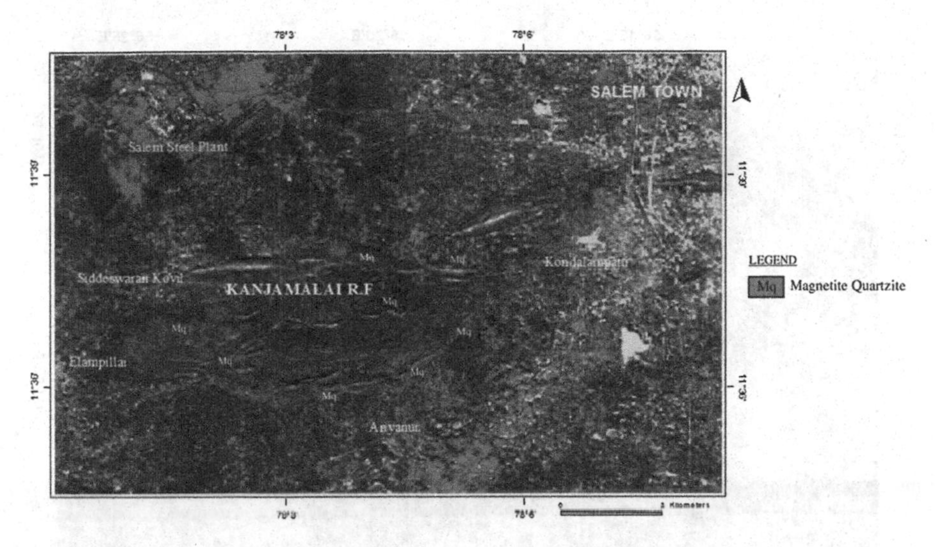

Fig. 4.23 Colour combination of band ratios ((1 + 3)/2, (3 + 5)/4, (5 + 7)/6) in *red*, *green* and *blue* (color figure online)

In case of iron oxide mineral exploration, the algorithm is relatively simple. Iron oxide minerals such as hematite, goethite and jarosite show an absorption trough in blue band and two reflection peaks at red and near-infrared parts of spectrum. Band 3/band1 ratio for landsat, band 2/band 1 ratio for IRS, SPOT and ASTER, can be used for enhancing these minerals. Other image processing methods such as PCA and SAM can also be used.

4.4.6 Remote Sensing Application for Chromite Exploration

Rajendran et al. (2012) have used ETM+ and ASTER data for chromite bearing mineralized zones in Semail ophiolite massifs of the northern Oman mountains. They used the capabilities of Landsat TM and Advanced Spaceborne Thermal Emission and Reflection Radiometer (ASTER) satellite data; applying image processing methods including non-correlated stretching, different band rationing and PCA for mapping chromite bearing areas. The study results show that the processed VNIR and SWIR spectral wavelength regions are promising in detecting the areas of potential chromite bearing mineralized zones within the ophiolite region, and proved to be successful for mapping of serpentinized harzburgite containing chromites. Figure 4.24 shows the colour combination of PC images that depicts several lithological units in the area.

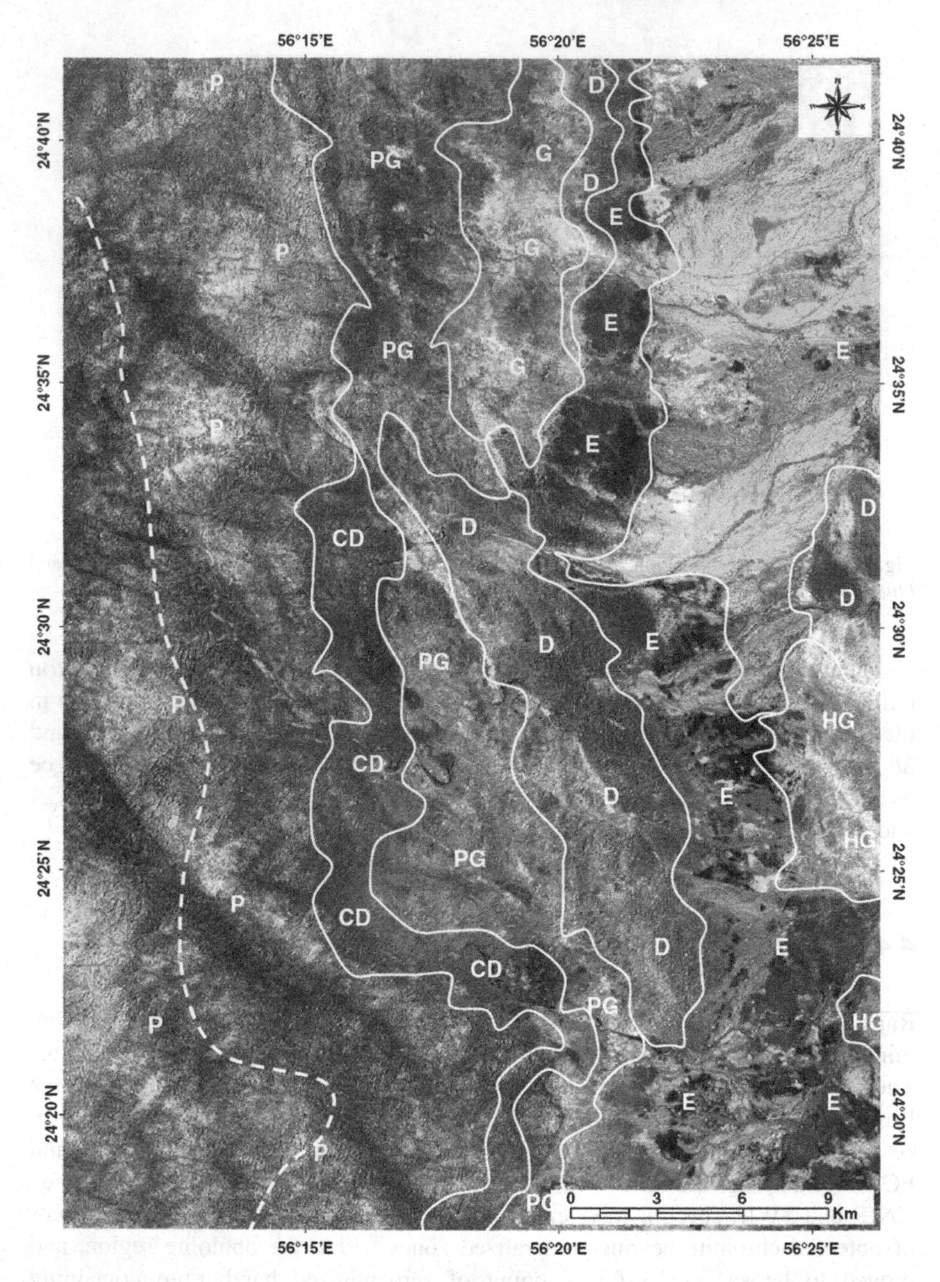

Fig. 4.24 Remote sensing image showing geology of the area; PC7, PC5 and PC4 of PCA bands. The index shows: *E* Basic extrusive rocks primarily spilites pillow lava, conglomerate; *D* Diabase dyke swarms; *G* Gabbro; *HG* Gabbroid hypabyssal rocks; *PG* Cumulate layered gabbro; *P* and *CD* Sheared serpentinized harzburgite (adopted from Rajendran et al. 2012)

Further Readings

Abrams MJ, Brown D, Lepley L, Sadowski R (1983) Remote sensing for porphyry copper deposits in southern Arizona. Econ Geol 78(4):591–604

Abrams M, Hook S, Ramachandran B (2002) ASTER user handbook (Ver. 2). Jet Propulsion Laboratory, California Institute of Technology, NASA, USA, p 135

Bernstein R (1983) Image geometry and rectification. In: Colwell RN (ed) Manual of remote sensing, 2nd edn. American Society of Photogrammetry, Falls Church, Virginia, pp 873–922

Chaves PS Jr, Berlin GL, Sowers LB (1982) Statistical method for selecting landsat MSS ratios. J Appl Photographic Eng 8:23–30

Campbell JB (2007) Introduction to remote sensing. The Guilford Press, New York, p 626

Carter WD, Rowen LC, Huntington JF (2013) Remote sensing in mineral exploration, Elsevier, Amsterdam, p. 186

Crosta AP, De Souza Filho CR, Azevedo F, Brodie C (2003) Targeting key alteration minerals in epithermal deposits in Patagonia, Argentina, using ASTER imagery and principal component analysis. Int J Remote Sens 24(21):4233–4240

Crosta A, Moore J (1990) Enhancement of Landsat Thematic Mapper imagery for residual soil mapping in SW Minas Gerais State, Brazil-A prospecting case history in greenstone belt terrain. In: Thematic conference on remote sensing for exploration geology—methods, integration, solutions, Calgary, Canada, pp 1173–1187

Di Tommaso I, Rubinstein N (2007) Hydrothermal alteration mapping using ASTER data in the Infiernillo porphyry deposit, Argentina. Ore Geol Rev 32(1):275–290

Dimitrijevic MD, Dimitrijevic MN, Djordjevic M, Vulovic D (1971) Geological map of the Pariz area, scale 1: 100,000. Geological Survey of Iran, Tehran

Drury SA, Hunt GA (1989) Geological uses of remotely-sensed reflected and emitted data of lateritized Archaean terrain in Western Australia. Int J Remote Sens 10(3):475–497

Drury SA (1993) Image interpretation in geology, 2nd edn. Allen and Unwin, London

Economic Geology (1983) An issue devoted to techniques and results of remote sensing. 78:573–770

Eklundh L, Singh A (1993) A comparative analysis of standardised and unstandardised principal components analysis in remote sensing. Int J Remote Sens 14(7):1359–1370

Fisher WA (1975) History of remote sensing. In: Reeves RG (ed) Manual of remote sensing. American Society Of Photogrametry, Falls Church, Virginia, pp 27–50

Galvao LS, Formaggio AR, Tisot DA (2005) Discrimination of sugarcane varieties in Southeastern Brazil with EO-1 Hyperion data. Remote Sens Environ 94(4):523–534

Govil H (2015) Identification and mapping of hydrothermally altered minerals in and around Askot base-metal mineralization of Kumoan Himalaya, India using EO-1 Hyperion data. Int J Adv Remote Sens GIS Geogr 3(1):1–8

Gupta RP (2008) Remote sensing geology. Springer, Berlin, p 655

Gupta RP (2013) Remote sensing geology. Springer Science & Business Media, Berlin

Honarmand M, Ranjbar H, Shahabpour J (2011) Application of spectral analysis in mapping hydrothermal alteration of the Northwestern Part of the Kerman Cenozoic Magmatic Arc, Iran, Journal of Sciences, Islamic Republic of Iran, 22(3): 221-238

Hubbard BE, Crowley JK (2005) Mineral mapping on the Chilean-Bolivian Altiplano using co-orbital ALI, ASTER and Hyperion imagery: data dimensionality issues and solutions. Remote Sens Environ 99(1):173–186

Hunt GR (1977) Spectral signatures of particulate minerals in the visible and near-infrared. Geophysics 42:501–513

Hunt GR, Ashley RP (1979) Spectra of altered rocks in the visible and near infrared. Econ Geol 74:1612–1629

Hunt GR, Salisbury JW (1970) Visible and near-infrared spectra of minerals and rocks: I silicate minerals. Modern Geol 1:283–300

Jensen JR (1996) Introductory digital image processing: a remote sensing approach. Prentice Hall, Upper Saddle River, NJ, p 7458

Kruse FA, Lefkoff AB, Boardman JW, Heidebrecht KB, Shapiro AT, Barloon PJ, Goetz AFH (1993) The spectral image processing system (SIPS)—interactive visualization and analysis of imaging spectrometer data. Remote Sens Environ 44(2):145–163

Kruse FA, Boardman JW, Huntington JF (2003) Comparison of airborne hyperspectral data and EO-1 Hyperion for mineral mapping. IEEE Trans Geosci Remote Sens 41(6):1388–1400

Lillesand TM, Kiefer RW (2003) Remote sensing and image interpretation. Wiley, New Jersey, p 722

Lillesand T, Kiefer RW, Chipman J (2014) Remote sensing and image interpretation. Wiley, New Jersey

Loughlin WP (1991) Principal component analysis for alteration mapping. Photogram Eng Remote Sens 57(9):1163–1169

Lowman PD (1969) Geologic orbital photography, experience from the Gemini program. Photogrametrica 24:77–106

Mars JC, Rowan LC (2006) Regional mapping of phyllic-and argillic-altered rocks in the Zagros magmatic arc, Iran, using Advanced Spaceborne Thermal Emission and Reflection Radiometer (ASTER) data and logical operator algorithms. Geosphere 2(3):161–186

Mather PM (1987) Computer processing of remotely-sensed images, an introduction. Wiley, New York, p 352

Navai I, Mehdizadeh-Tehrani S (1994) Alteration mapping by remote sensing techniques in south Iran, a case study. In: Proceedings of the 15th Asian conference on remote sensing, Bangalore, India

Navalgund RR, Jayaraman V, Roy PS (2007) Remote sensing applications: an overview. Curr Sci 93(12):1747–1766

Niblack W (1986) An introduction to digital image processing. Strandberg Publishing Company

Rajendran S, Thirunavukkarasu A, Balamurugan G, Shankar K (2011) Discrimination of iron ore deposits of granulite terrain of Southern Peninsular India using ASTER data. J Asian Earth Sci 41(1):99–106

Rajendran S, Al-Khirbash S, Pracejus B, Nasir S, Al-Abri AH, Kusky TM, Ghulam A (2012) ASTER detection of chromite bearing mineralized zones in Semail Ophiolite Massifs of the northern Oman Mountains: exploration strategy. Ore Geol Rev 44:121–135

Ranjbar H (2011) Remote sensing, applications to geophysics. In: Encyclopedia of solid earth geophysics. Springer, Netherlands, pp 1035–1039

Ranjbar H, Honarmand M (2007) Exploration for base metal mineralization in the southern part of the Central Iranian Volcanic Belt by using ASTER and ETM+ data. J Eng Sci 3:23–34

Ranjbar H, Roonwal GS (2002) Digital image processing for lithological and alteration mapping, using spot multispectral data. A case study of Pariz area, Kerman Province. In: Proceedings of SPIE, remote sensing for environmental monitoring, GIS applications, and geology, Iran, vol 4545

Rowan LC, Mars JC (2003) Lithologic mapping in the Mountain Pass, California area using advanced spaceborne thermal emission and reflection radiometer (ASTER) data. Remote Sens Environ 84(3):350–366

Rowan LC, Schmidt RG, Mars JC (2006) Distribution of hydrothermally altered rocks in the Reko Diq, Pakistan mineralized area based on spectral analysis of ASTER data. Remote Sens Environ 104(1):74–87

Sabins FF (1999) Remote sensing for mineral exploration. Ore Geol Rev 14(3):157–183

Sanjeevi S (2008) Targeting limestone and bauxite deposits in southern India by spectral unmixing of hyperspectral image data. Int Arch Photogram Remote Sens Spat Inf Sci 37:1189–1194

Shahriari H, Ranjbar H, Honarmand M (2013) Image segmentation for hydrothermal alteration mapping using PCA and concentration–area fractal model. Nat Resour Res 22(3):191–206

Tangestani MH, Moore F (2001) Porphyry copper potential mapping using the weights-of-evidence model in a GIS, northern Shahr-e-Babak, Iran. Aust J Earth Sci 48 (5):695–701

Tangestani MH, Moore F (2002) Porphyry copper alteration mapping at the Meiduk area. Iran. International Journal of Remote Sensing 23(22):4815–4825

Tangestani MH, Mazhari N, Agar B, Moore F (2008) Evaluating Advanced Spaceborne Thermal Emission and Reflection Radiometer (ASTER) data for alteration zone enhancement in a semi-arid area, northern Shahr-e-Babak, SE Iran. Int J Remote Sens 29(10):2833–2850

Van der Meer FD, Van der Werff HM, van Ruitenbeek FJ, Hecker CA, Bakker WH, Noomen MF, Woldai T (2012) Multi-and hyperspectral geologic remote sensing: A review. Int J Appl Earth Obs Geoinf 14(1):112–128

Zhang X, Pazner M, Duke N (2007) Lithologic and mineral information extraction for gold exploration using ASTER data in the south Chocolate Mountains (California). ISPRS J Photogram Remote Sens 62(4):271–282

Chapter 5
Survey in Exploration

5.1 Aims and Objectives of Surveying in Exploration

Survey is basic for geological mapping on the surface as well as underground. Surveying as adopted for generating a topographical map, underground plan or vertical section on which geological data are plotted, and for marking on the map, plan or section. They are used for measuring the volumes of prospecting pits and trenches and the minerals or ores recovered, for determining the recovery, swell and tonnage factors. The location and orientation of exploratory boreholes and drivages are computed on the basis of surveying. To the geologist, a knowledge of principles of the different methods of surveying helps to effectively plan exploration. Therefore, surveying aims to carry out measurements for representing a point or set of points on the surface of the earth, in their correct horizontal and vertical relationship. The measurements may be wholly linear or a combination of linear and angular, depending upon the methods of surveying adopted. For purposes of limiting and localizing minor errors and to prevent their accumulation, surveying is always carried out from the whole to the part. In geological mapping of a mineral project for example, a set of main control points is established at the periphery of the prospect and connected with one or more permanent reference stations like Survey of India triangulation stations or state boundary pillars or revenue tri-junction posts. If the area is large, another set of control points is set up within the prospecting lease and connected to the main set of control points. Topographic and geological details are plotted with reference to the second set of control points. Thus, the operation will involve three stages; viz. main, subsidiary and detailed survey.

Table 5.1 gives a summary of the different surveying instruments and the quality of accuracy in data given. These are broad guidelines intended to help in material selection of survey procedure to be adopted. To learn more on survey, one may refer to special books on the subject. Table 5.2 gives recommended survey methods

G.S. Roonwal, *Mineral Exploration: Practical Application*,
Springer Geology, DOI 10.1007/978-981-10-5604-8_5

Table 5.1 Survey type and equipments with reference to the accuracy needed

1.	Chain survey	It is employed when no angle measuring instrument is available. It can be carried out with the help of untrained hands. The country should be more or less flat. The method can also be employed for the interior filling of theodolite surveys
2.	Compass survey	It is good for preliminary work. Distances are, in general, obtained by taping or pacing. As such it cannot be employed for accurate work. It can be used for interior filling of chain surveys and for normal geological mapping, on large scale. It can also be used in stope surveys underground (with taping)
3.	Plane table survey	It is useful in flat or gently undulating topography. It is as accurate as chain surveys, to be usefully employed for interior filling of theodolite surveys. When telescopic alidade, equipped with Beaman stadia arc is available, the method is applied even in fairly rugged terrain. It is recommended for, detailed mapping and proving of mineral deposits—underground mine surveys, and in geochemical and geophysical surveys
4.	Theodolite survey	Traverses are required for accurate work—when high accuracy is needed theodolite triangulation is recommended. Tacheometric surveys are as accurate as chain surveys, but the accuracy is relatively higher in rugged country. The method is useful in contouring hilly country. It is useful in proving mineral deposits, mines surveys and correlation surveys
5.	Abney level	Useful in contouring in conjunction with compass surveys or in the interpolation of heights in such surveys in rugged country. It can be used also with chain surveys
6.	Dumpy level	Useful in accurate contouring of theodolite and plane table surveys in a flat or gently undulating topography

in exploration of different types of mineral deposits. Table 5.1 gives recommended survey methods in exploration of different mineral deposits.

The choice will depend upon the following:

(i) For the precious or valuable metals, a more accurate method of survey is adopted.
(ii) Selection of survey method is adopted considering if a preliminary or a detailed investigation is planned.

5.2 Methods of Surveying

Horizontal control, relative distance between the control points as measured on a horizontal plane, is established by triangulation and closed theodolite traversing for the main surveys. Traversing, tachometry, plane tabling and compass and tape surveying are used to establish horizontal control for the subsidiary surveys and detailed surveys carried out within the triangulation network or closed theodolite traverse.

Table 5.2 Showing few examples of mineral deposits and the methods of survey

	Type of nature of deposit	Instruments or methods to be employed
1.	Limestone and clay in a flat country	If high order of accuracy is not required, the compass and pacing with Abney level will do. If a higher order of accuracy is needed, plane table with dumpy level is necessary
2.	Limestone or coal in flat country	Theodolite traverse with dumpy level will be required
3.	Metalliferous lode in plane country (gold or copper or manganese, etc.)	Theodolite traverse, or plane table or chain with dumpy level will be required
4.	Metalliferous lode in a hilly country	Tacheometer survey (or Telescopic alidade with Beaman stadia arc) may be employed
5.	Iron ore occurring as a capping on a flat topped hill	Theodilite, traverse with tacheometer and levelling by stadia or dumpy level gives satisfactory results. Compass survey should avoid results
6.	Mine survey for coal or metal	Theodolite traversing or dial survey may be done. Underground traverses may also be done with plane table with Beaman stadia arc attachment
7.	Stope survey	Compass and tape survey (Brunton compass is convenient) is best suited

5.2.1 *Triangulation*

In triangulation, horizontal control is established by a system of interconnected triangles in which the length of only one side, called the base line, and all the angles of the triangles are measured precisely. The lengths of the remaining lines (sides of triangles) are known as triangulation stations and the network of triangles is called a triangulation system or triangulation. Based on aim of the survey, the area covered and the accuracy of the measurement desired, the triangulation systems are classified as shown in Table 5.3. The maximum area being statutorily limited to less than 25 km^2 for a prospecting lease, and 10 km^2 for a mining lease. The triangulations carried out for surveying such areas belong to the tertiary category of triangulations. Here, precise surveying procedures normally used for primary/secondary triangulations are not required the purpose.

Principles of Triangulation

A triangulation survey in exploration is carried out based on design of network of triangles need to be decided based on the shape of the property. When the areas are small, a well-connected polygonal shape of the network with a central station known as a "Hub station", is preferred to an open-chain network of triangles. Triangles selected need to be well conditioned, the angles should not be less than

Table 5.3 Classification of triangulation systems

S.L.	System	Primary triangulation	Secondary triangulation	Tertiary triangulation
1.	Average triangle closure	Less than 1 s	3 s	6 s
2.	Maximum triangle closure	Note more than 3 s	8 s	12 s
3.	Length of base line	5–15 km	1.5–5 km	0.5–3 km
4.	Length of the sides of triangles	30–150 km	8–65 km	1.5–10 km
5.	Actual error of base	1 in 300,000	1 in 150,000	1 in 75,000
6.	Probable error of base	1 in 1,000,000	1 in 500,000	1 in 250,000
7.	Discrepancy between two measures of a section	10 mm/km	20 mm/km	25 mm/km
8.	Probable error of computed distance	1 in 60,000 to 1 in 250,000	1 in 20,000 to 1 in 50,000	1 in 5000 to 1 in 20,000
9.	Probable error in astronomical azimuth	0.5 s	2 s	5 s
10.	Applicability	Geodetic surveying, i.e. for mapping a whole country or to furnish most precise control points to which secondary triangulation may be connected	Establishing closer control points within a primary network or main system of control for smaller areas	For precise control network within the primary and secondary networks from which subsidiary surveys for location details can be carried out

30° and more than 120°. Triangulation involves measurement of all angles of the triangles. (i) measurement of only one line, known as “Base Line” is done, and the selection of the base line should be done carefully. (ii) Another line in the network is selected and measured accurately to check the accuracy of the baseline, it is called “Check-Base”. Both “Base” and “Check-Base” should be on a flat and level ground; straight stretch of rail-track or highway is ideal. The check-base should be as far away from the base line as possible, and nearly at right angle to the direction of the base line. The base line may be 500–1000 m in length. The check-base need not be as long as the base line.

5.2.2 Traversing

In traversing, horizontal control is established by measuring the lengths and directions of the lines joining the traverse stations. Traverses are of two types:

1. **Closed traverse**: A closed traverse closes upon the starting station (usually a triangulation station, a tri-junction point or other permanent control point) from which the traverse is commenced. It is called a polygon closed traverse to distinguish it from one which is closed between two previously established control points. The accuracy of work in the case of polygon closed traverse can be judged by comparing the initial and final bearings of the closing line, and the initial and closing coordinates of the survey. In the case of a traverse which is closed between two previously established control points, the accuracy of the traverse is checked by the agreement between the bearings and the coordinates of the line joining them obtained from the traversing, with the corresponding values obtained from the more accurate original survey. In exploration surveys, closed traversing is adopted to establish the main horizontal control points when the area to be covered is flat or when it is not very large, and the distances between the stations can be measured without difficulty. It is used to establish additional or subsidiary horizontal control points for detailed surveys between the permanent reference points established earlier by triangulation.

   Traversing involves the following operations:

(a) Reconnaissance and setting up traverse stations: The aim of reconnaissance is to adopt the shortest traverse route, which offers best ground for taping and avoids short legs. The site at which traverse stations are located must be stable; unlikely to be affected by mining or natural agencies. Here, a sketch is made of the route to be followed, the location of the stations and the places where distinct changes of slopes occur along each leg. The permanent stations are marked in the same way as the triangulation stations.

(b) Measurement of angles and distances: A theodolite is used for measuring the angles. A microptic theodolite with optical plumbing will make rapid progress without comprise in accuracy. The instrument is accurately levelled and centred on the traverse stations with telescope in the face-left position. The telescope is directed towards the back station and is sighted, and plates are clamped. Accurate bisection is affected by tangent screws. Initial reading of the horizontal circle is recorded. The horizontal circle may be set to read zero. Now, the upper plate clamp is released and the telescope is directed towards the forestation and the clamp is tightened. The station is accurately bisected with the help of upper plate tangent screw, and the reading is taken. The difference of the initial and the final reading will give the horizontal angle between back and forestations. The process is repeated in the face-right position, also for the sake of accuracy, two sets of readings are taken at each setting. The angle of elevation or depression (vertical angles) is also measured and recorded.

Calculating of bearings and coordinates, the azimuth (i.e. the clockwise angle of a line with the true north of the first line) is either already known or determined. The external angle of each of the subsequent traverse lines is converted into a whole circle or azimuth bearing. In a closed traverse, initial and closing azimuths of the first line should agree.

2. **Open traverses**: Open traverses are adopted to provide subsidiary horizontal control points, from which details can be filled up. Open traverses normally do not require the same refinement or care as the closed traverses, which partly replace triangulation. Horizontal angles are measured only once or rarely twice on each face. In prospecting and exploration, open traverses are carried out for fixing the position of the tacheometer or plane table stations for filling in details. The traverse stations may also be used as end points for taps and compass traverses from which offsets are taken to the various surface and geological features (Table 5.4).

5.2.3 Levelling

Primary vertical control points to which different kinds of surveys are normally referred to are the "primary protected bench marks" established by the Survey of India on the basis of precision levelling. The mean sea level determined on the basis of tidal observations at selected sea ports forms the datum for these benchmarks. The central and state public works or civil engineering department, railways, city and town planning authorities establish local benchmarks connected to these via secondary and tertiary benchmarks.

Spirit levelling: Spirit levelling consists of taking measurements with a levelling instrument and a graduated staff. Usually a dumpy level is used for this purpose.

In surveys carried out for exploration, one or more benchmarks are established within the area being explored, by carrying fly-levelling from the nearest Survey of India branch within a reasonable distance, and the assumed datum is used for the purpose. The benchmarks so established should be sited on firm rock not likely to be disturbed by natural or human agencies. Benchmarks established in unconsolidated ground are unreliable, especially when subjected to vibrations due to mining operations, movement of heavy vehicles or railways.

Table 5.4 Method of booking—measurement of angles in traversing

Traverse	From T to T1	To No. 1	Shaft	Observer booker	P.Q. M.N.	Instrument	Microptic theodolite with optical plumbing Inst. No. Th-2			
To station	Horizontal angles at station 6					Vertical angles from 6 to 7				Changes of slopes
	F.L.	Included angle	F.R.	Included angle	Mean angle	Face	Reading	Vertical angle	Mean	
							6 to a			
5	273-55-55		93-27-31	170-22-34	170-22-36	L	359-32-32	00-27-28		
7	84-17-35	170-22-40	263-50-05			R	180-28-10	00-28-10	00-27-49	
5	07-37-30		190-07-56			L	3-28-10	3-28-10		
7	178-00-08	170-22-38	00-30-28	170-22-32		R	176-31-36	3-28-24	3-28-17	
						L	357-24-05	2-35-55		
						R	182-36-28	2-36-28	2-36-12	
	Horizontal angles at station 6					Vertical angles from 6 to 7				
							7 to a			
6	318-41-10		297-10-31			L	358-41-16	01-18-44		
8	163-06-50	204-25-40	141-36-05	204-25-34		R	181-19-03	01-19-03	01-18-54	
					204-25-35		b to a			
6	36-21-58		02-05-48			L	357-12-24	02-47-36		
8	240-47-31	204-25-30	206-31-21	204-25-33		R	182-47-44	02-47-44	02-47-40	
							b to 8			
						L	355-14-42	04-45-18		
						R	184-45-40	04-45-40	04-45-29	

The modern Engineer's level has replaced the dumpy level, which was earlier used for ordinary levelling jobs. However, as the sturdy dumpy level is still used in some mines, it is briefly described here. It consists of a telescope, whose barrel is cast solid with a vertical spindle carried on a triangular plate (tribrach). The tribrach is supported on a bottom plate or trivet stage by means of three levelling screws. A sensitive spirit level is mounted on the telescope by means of a hinge at one end and a capstan screw at the other end. The capstan screw is meant for making the axis of the spirit level parallel to the optical axis of the telescope. A smaller spirit level is mounted at right angles to the main spirit level near the eyepiece of the telescope. The trivet state incorporates a quick levelling device, and can be readily mounted on a telescopic tripod. When correctly adjusted, the axis of the spirit level is parallel to the optical axis of the telescope and both the axes are perpendicular to the vertical axis. The line of sight is horizontal for all positions of the telescope, if the main spirit level remains centralized in all positions. This is achieved by centralizing the main spirit level in two directions at right angles to each other. In practice, however due to wind pressure, a slight sinking of the tripod or movement caused by the activities of the observer can disturb the verticality of the axis of rotation of the dumpy level and the bubble may not be quite central for a given sight. The bubble can be recentralized by means of one of the foot-screws.

In the modern Engineer's tilting levels, the disadvantages of the dumpy level have been overcome by pivoting the telescope at its junction with the vertical axis, so that it can be revealed for each line of sight by means of a fine levelling screw, without affecting the height of the line of sight. A small circular spirit level is provided for quick preliminary levelling at each setting. The images of the two ends of the main spirit bubble are brought together by a specially designed optical system placed alongside the eyepiece of the telescope. This makes it possible for the surveyor to centralize the bubble by coincidence of the images. Further, one can check the coincidence of the bubble as the staff reading is taken. A modern tilting level must satisfy the following conditions if it is correctly adjusted: (a) the axis of rotation must be approximately vertical when the circular spirit level is central, (b) the main spirit level must be central when the line of collimation is horizontal and (c) the fine levelling screw should be at zero when the main spirit level bubble is central and the axis of rotation is truly vertical. Procedures for testing and correcting these adjustments can be found in the textbooks on surveying listed in the references given at the end of this chapter.

The levelling stages consist of 3–4 sections, either hinged or telescopically connected, and made of seasoned wood. The stages are graduated in metres, decimetres and centimetres as shown in Fig. 5.1.

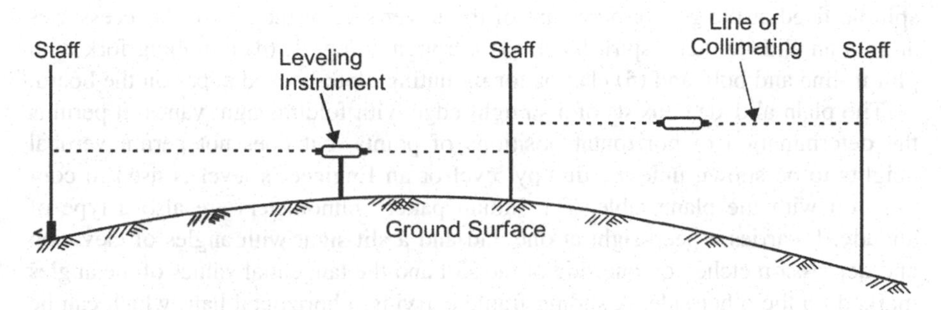

Fig. 5.1 Methods of levelling

5.2.4 Contouring

Of the different methods in representing relief (i.e. surface features) in a map, the most convenient method is by contours. A contour line can be defined as the intersection of a level plain and the ground surface. Every point on a contour is, therefore, at the same elevation. The difference between the elevations of points on two successive contours is called the contour interval. The contour interval for any particular plan is selected by taking into account the purpose of the survey, the extent of the survey, the nature of the country and the scale of the plan.

5.2.5 Tacheometry

Tacheometry is a method of rapidly measuring horizontal and vertical distances exclusively by instrumental observations. It does not involve any linear measurement, and in most cases, it is possible to make the observations for a number of points from one fixed station. It is used for expeditiously plotting contours, topographical details and geological features within the horizontal control framework, provided by the principal and subsidiary surveys. It is particularly suited for detailed mapping of a rough country where linear measurement by chain or tape is slow and inaccurate and ordinary levelling is tedious.

5.2.6 Plane Tabling

Plane tabling is the easiest and only method of plotting of topographical and geological details directly in the field. The plane table is a rectangular (65 × 45 cm) or square (65 × 65 cm) drawing board made of well-seasoned wood. It is attached to a tripod with a quick levelling head by means of a vertical spindle fixed at the geometric centre of the lower side of the board. Its accessories are (1) an alidade, (2) a spirit level, (3) a trough compass, (4) plumbing fork with plumb-line and bob, and (5) clamps for mounting cloth-backed paper on the board.

The plain alidade consists of a straight edge with folding sight vanes. It permits the determination of horizontal positions of points, but does not permit vertical heights to be shown unless a dumpy level or an Engineer's level is used in conjunction with the plane table. The Indian pattern clinometers are also a type of alidade. It carries a peep-sight at one end and a slit-sight with angles of elevation and depression etched on one side of the slot and the tangential values of the angles marked on the other side. A sliding frame carrying a horizontal hair, which can be clamped in any desired position with a thumb screw is provided on the slit-sight to facilitate reading. The sight vanes are hinged to an upper plate, which can be levelled in relation to the base plate by means of a levelling screw and a small spirit level. When levelled correctly, the zero on the slit-sight is at the same level as the peep-hole. The difference of level between the point being sighted and the plane table is found by multiplying the tangent scale reading by the distance of the point. A telescopic alidade consists of a tacheometric telescope aligned with the straight edge. It carries a graduated vertical circle for taking inclined sights. It is used in conjunction with a stadia rod or levelling staff. A spirit level is used for levelling the table. A compass is used in some surveys for aligning the plane table with the magnetic meridian with the help of a north line drawn on the paper. The plumbing fork is used for marking a point corresponding to the instrument station on the paper mounted on the plane table.

Methods of Surveying with plane table:

(1) Radiation: In this method, the plane table is set up on a control point from which the points to be plotted are visible. The table is clamped and the control point of station is transferred to the paper by the plumbing fork. Distance from the station to each point is carefully measured and scaled off on the corresponding ray as shown in Fig. 5.2. The method is used only for large-scale work.

(2) Traversing: It is similar in principle to close traversing with a theodolite. The plane table is successively set up over previously selected traverse stations, and back and forward sights are taken and lines are drawn at each setting. The distances are measured by tape and plotted to scale to locate the stations. Accuracy of the traverse is checked by the coincidence of the starting station, as initially plotted, with the intersection of the forward sight from the last station. This method is useful for surveying roads, rivers, boundaries.

(3) Intersection: Intersection is the method commonly used for plotting the details. The method consists of setting up the plane table first at one end of a base line of known length. The base line is drawn to scale on the map sheet after transferring the instrument station to the paper by the plumb bob and sighting the other end of the base line with the alidade. Rays are then drawn to the points to be plotted. The plane table is then shifted to the other end of the base line and clamped after being aligned by taking a back sight along the plotted base line.

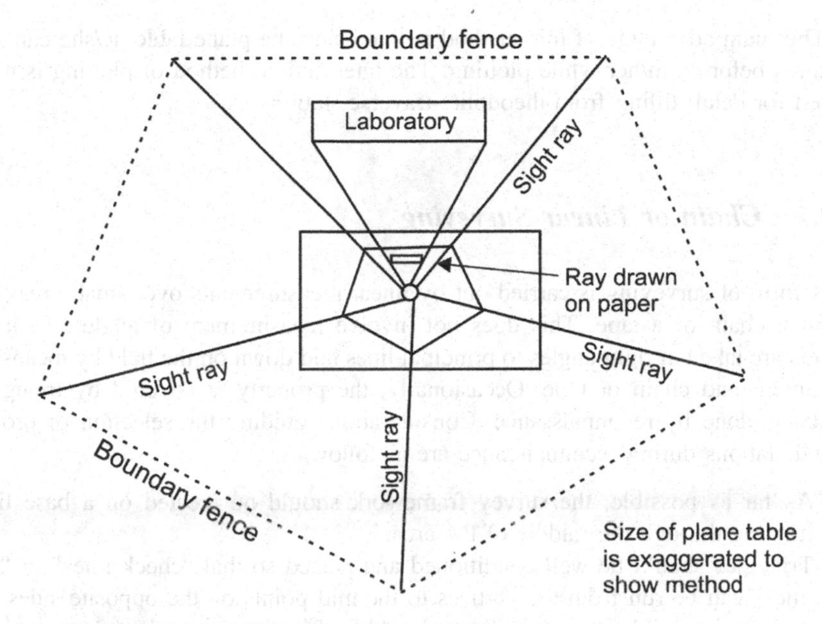

Fig. 5.2 Schematic diagram of the plane table—radiation pattern

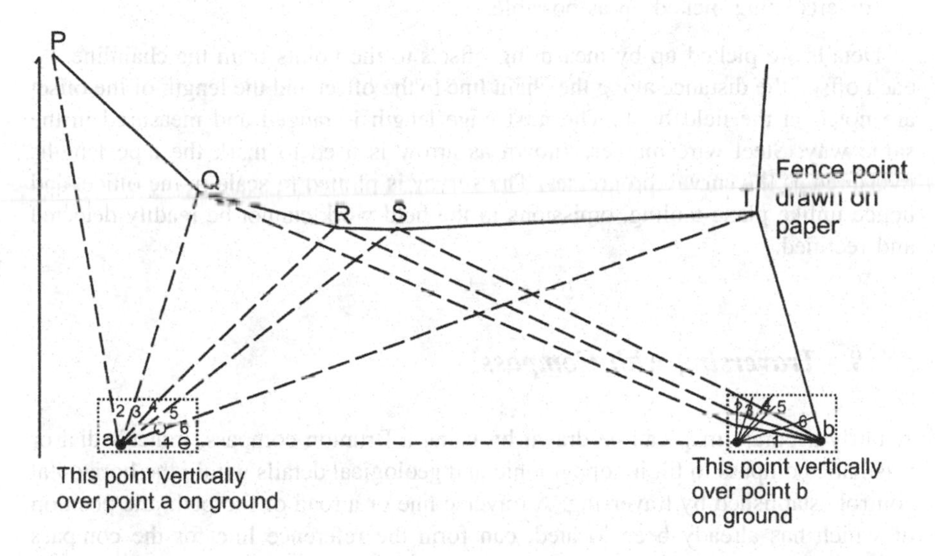

Fig. 5.3 Schematic diagram of the plane table—intersection pattern

Rays are now drawn to the points to be plotted intersecting the corresponding rays drawn from the previous plane table setting. Topographical and geological details can be sketched with the help of these intersection points. This is shown in Fig. 5.3.

The main advantage of this method is that using the plane table he/she can see features before him/her while plotting. The intersection method of plotting is well suited for detail filling from theodolite-traverse stations.

5.2.7 Chain or Linear Surveying

This form of surveying is carried out by linear measurements over small areas by using a chain or a tape. This does not involve measurement of angles. Usually offsets are taken at right angles to principal lines laid down on the field by means of alignment and chain or tape. Occasionally, the property is covered by triangles which is done by reconnaissance. Considerations guiding the selection of provisional stations during reconnaissance are as follows:

(a) As far as possible, the survey framework should be erected on a base line running through the middle of the area.
(b) Triangles should be well conditioned and placed so that "check lines" or "tie lines" can be run from the vertices to the mid-points on the opposite sides.
(c) As far as possible, the chain lines should be free from any obstacles.
(d) Survey lines from which offsets are taken should run as close to the surface details being picked up as possible.

Details are picked up by measuring offsets to the points from the chainline. At each offset, the distance along the chain line to the offset and the length of the offset are noted in the field book. The next page length is ranged and measured in the same way. Steel wire marker, known as arrow is used to mark the tape lengths measured as the survey progresses. The survey is plotted to scale in the office, and hence unlike plane tabling, omissions in the field work cannot be readily detected and rectified.

5.2.8 Traversing with Compass

A fairly accurate map can be drawn by using a Brunton compass, miner's dial or prismatic compass to fill in topographic and geological details within the horizontal control established by traversing. A traverse line or a road or footpath, the position of which has already been located, can form the reference line for the compass survey. Magnetic bearings of details or objects are taken from the reference line, and the respective distances are measured by pacing, taping, etc. Care is taken to ensure that at least two rays are taken to each point of the detail so that it can be plotted by intersection. This method is faster than chain surveying. Further, points which are at a distance can be picked up by intersection. The date of traverse should be noted, in order to check the magnetic variations. Care also is to be taken to see that no local attractions are existing in the vicinity of the survey lines.

5.3 Differential Global Positioning System (DGPS)

Concept and Methodology

Differential GPS is an enhancement to Global Positioning System that provides improved location accuracy, from the 15-m nominal GPS accuracy to about 10 cm in case of the best implementations. DGPS uses a network of fixed, ground-based reference stations to broadcast the difference between the positions indicated by the GPS (satellite) systems and the known fixed positions. These stations broadcast the difference between the measured satellite pseudoranges (https://en.wikipedia.org/wiki/Pseudorange) and actual (internally computed) pseudoranges, and receiver stations may correct their pseudoranges by the same amount. The digital correction signal is typically broadcast locally over ground-based transmitters of shorter range. The procedure is very simple. A GPS antenna is fixed at a local station, whose coordinates and elevation are known accurately. The second antenna, which is connected to the first one without a wire, moves in the field and collects data such as elevation, lithological contacts, bedding plane measurements, etc. Accurate geological and topographical maps can be prepared rapidly.

Surveying by Total Station Theodolite

A total station theodolite is an electronic/optical instrument, which is used in modern surveying (https://en.wikipedia.org/wiki/Surveying) and building construction (https://en.wikipedia.org/wiki/Construction). The total station is an electronic theodolite (https://en.wikipedia.org/wiki/Theodolite) (transit) integrated with an electronic distance metre (https://en.wikipedia.org/wiki/Distance) (EDM) to read slope distances from the instrument to a particular point. Robotic (https://en.wikipedia.org/wiki/Robotics) total stations allow the operator to control the instrument from a distance via remote control. This eliminates the need for an assistant staff member as the operator holds the reflector and controls the total station from the observed point. The equipment is equipped with a GPS. The working procedure is simpler than a normal theodolite. The geologist can just walk on the boundary of lithologies and point the reflector towards the total station. It has a data card that can save the data. The data can be easily transferred to a computer and the maps are drawn using tailor-made softwares.

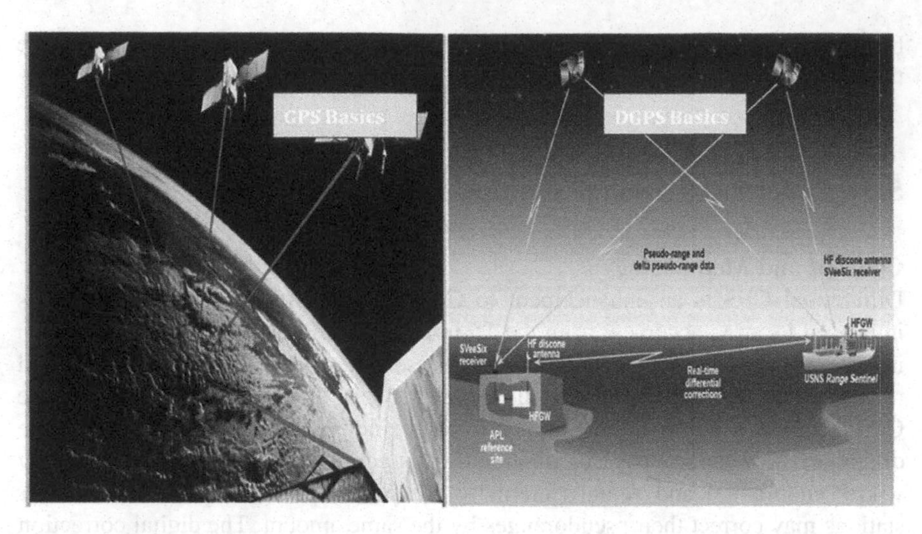

Fig. 5.4 DGPS and GPS navigation concept

5.3.1 DGPS Basic Concepts

Differential Global Positioning System (DGPS) is an enhancement to Global Positioning System, which is a satellite-based system that uses a constellation of 24 satellites and ground-based reference station to broadcast the difference between the positions indicated by the satellite system and takes the known fixed positions as shown in Fig. 5.4.

5.3.2 Purpose of the DGPS Survey

The main objective of the DGPS survey is for Satellite Image Geo-Metric Correction/Geo-Referencing.

5.3.3 Instrument and Technical Specification

Model	Leica GPS1200+
GNSS technology	SmartTrack
Channels	16 L1 + 16 L2
GPS1200+ receivers	GX1210+
Ports	1 power/controller port, Bluetooth® Wireless-Technology

Supply voltage	Nominal 12 VDC
Consumption	4.6 W receiver + controller + antenna
Standard antenna	Smart Track+ ATX1230+ GNSS
Post-processing with Horizontal	10 mm + 1 ppm, kinematic
Leica Geo Office Vertical	20 mm + 1 ppm, kinematic
Software Horizontal	5 mm + 0.5 ppm, static
All GPS1200+ Vertical	10 mm + 0.5 ppm, static
Horizontal	3 mm + 0.5 ppm, static
Vertical	6 mm + 0.5 ppm, static

Notes on performance: Figures quoted are for normal to favourable and accurate conditions. Performance and accuracies can vary depending on number of satellites, satellite geometry, observation time, ephemeris, ionosphere, multipath, etc.

5.3.4 DGPS Survey Steps

(a) GCP Marking/Identification on satellite Image.
(b) Identification of GCP's on the Ground.
(c) Base Establishment.
(d) GCP Coordinates Collection Using Leica Rover.
(e) 4 Photographs and Sketch of Each Photograph.
(f) Data Import and Analysis of Coordinates.
(g) Geo-Referencing of Satellite Image with the Help of Erdas Imagine9.2 Software.

5.3.5 DGPS Point Selection Criteria

(a) Set of Ground Control Points at different proposed locations.
(b) Ground Control Points are chosen such that it will represent Point feature on a Satellite imagery such as end point of any linear feature, corner points of a fence.
(c) DGPS Base Station should be placed on an elevated area for maximum satellite signal probability.
(d) DGPS Base station should be far from water bodies, tree canopy and building shadow.

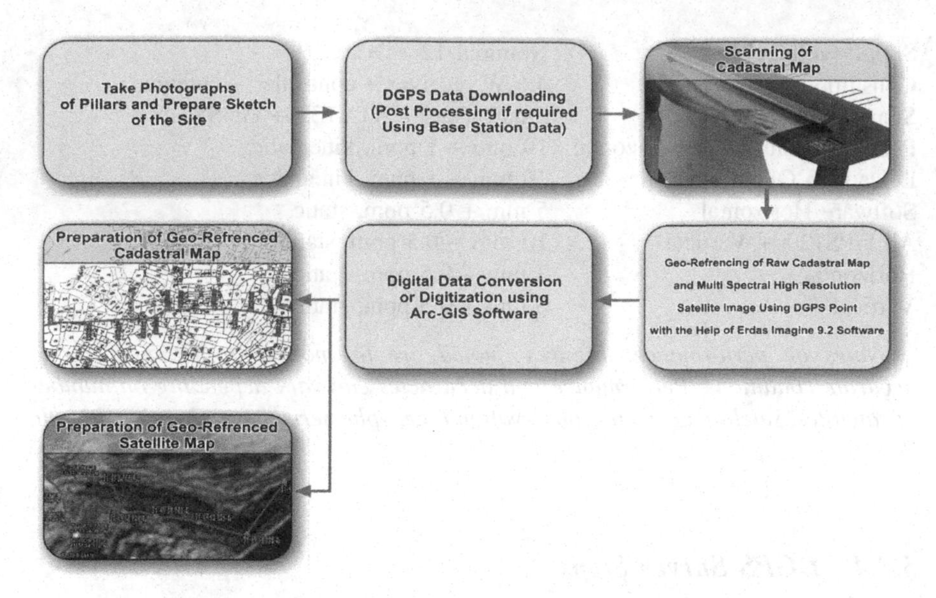

Fig. 5.5 Procedure of DGPS Surveys: Various companies' DGPS are available in market, but Leica Geosystem and Trimble are the best instruments; The cost of DGPS instrument is dependent upon horizontal and vertical accuracy

(e) Observation time should be at least 1 h for better accuracy. If observation time is more then accuracy will be high.
(f) For DGPS point collection, Satellites should be more then 6. If satellite availability is more then accuracy will be high.

These are shown in Fig. 5.5 and are listed below:

(a) DGPS Base Station should be placed on elevated area for maximum satellite signal probability.
(b) DGPS Base station should be far from water bodies, tree canopy and Building shadow.
(c) Observation time should be at least 2 h for better accuracy. If Time observation is more then accuracy will be high.
(d) For DGPS point collection, Satellites should be more then 6. If satellite availability is more then accuracy will be high.

Precautions during survey:

(a) Mining lease boundary Should be Latest and Updated on Cadastral Map.
(b) Base station should be Established 2 hours before starting the work for better accuracy.
(c) Observation time should not be less than 2 hours.

(d) The person who is familiar with Mining Lease Boundary should be with the Survey Team, so that he can help in GCP Identification on Ground.
(e) If there are any encroachment or changes in mining lease area, in this case, survey team should be aware of the situation.

5.4 Underground Surveys

The principles of surveying in underground workings are basically similar to those of surface surveys, but the methods used are dictated by the constraints imposed by poor lighting conditions, low height, limited space, steep slopes, constant activity connected with mining operations. Traversing by theodolite, and rarely by a miner's dial, is the most common method of underground survey. Special techniques and accessories are available for dealing with the difficulties encountered in carrying out a theodolite-traverse underground.

(1) Traverse stations: The traverse stations are usually fixed on the roof of the underground workings as pegs or markers fixed on the floor are liable to be dislodged by the movement of men and machines. The station is marked by driving a dry wooden plug into a hole drilling in the roof and a special brass spud is fixed in the plug, so that it projects only slightly below the roof.
(2) Centring of the theodolite is done by suspending a plumb bob from a hole in the spud. The theodolites used underground have a centring pin on top of the telescope for this purpose.
(3) For sighting the plumb-line, it is necessary to plane an illuminated cloth or paper screen behind it. A cap lamp or torch can be used for this purpose.
(4) For surveying steep workings in which the tripod cannot be set up, it may be necessary to mount the theodolite on a stretcher bar. It consists of two light strong alloy steel telescopic tubes which can be clamped by screw action between walls. The telescope is carried at one end of a bar, with counter-weight at the other end. This bar is mounted on the stretcher bar in such a way that it can be moved or turned in any direction and clamped in any desired position.
(5) Accessories for the theodolite in underground surveys:
    (i) Diagonal eyepiece—It is used for taking steep upward sights in which the target cannot be conveniently viewed through the eyepiece of the telescope.
    (ii) Auxiliary telescope—This is a smaller telescope mounted above or on the side of the normal telescope. It is used for taking steep downward sights.
    (iii) Strident level—This is a level which can be mounted on the trunnion axis and is used for ensuring that this axis is truly horizontal.
    (iv) Modern theodolites are equipped with battery-powered illumination systems for lighting up the vertical and horizontal circles if necessary.

(6) In underground levelling work, the staff is generally held in an inverted position with its base on the roof, so that height of the station in roof above the line of collimation is measured, in order to keep the reduced levels always positive. The datum to which the underground levels are referred is assumed and is marked on a benchmark at the lowest point within the mine.
(7) In steep workings, trigonometric levelling is used.
(8) Correlation or orientation of underground survey with the surface surveys is essential. The methods used for this purpose are the following:

   (a) Direct traversing via mine entry: This is possible when the underground workings are accessible through an adit or incline.
   (b) Shaft plumbing: The methods listed below are used where one or more vertical shafts provide access to the underground workings.

      (I) On wire in each shaft: This method is used when two shafts connected by the underground workings are available. At the surface, the true bearing of the line joining the plumb wires suspended in the two shafts is determined by connecting the wires directly or indirectly to triangulation stations. An underground traverse is run between the wires and their coordinates calculated by reference to an assumed meridian. The bearing of the plumb plane calculated from these coordinates is compared with the true bearing observed on the surface. If there is variation between the true bearing and the computed bearing, correction has to be applied to the computed bearing and also to all the computed bearing and also to all the underground bearings by the amount of variation.
      (II) Two or more wires in a single shaft: In this method, two wires are vertically suspended in the shaft to form a plumb plane. The azimuth of this plumb plane is determined and transferred on the reference base below ground. Two or more plumb planes may be formed by suspending a number of pairs of plumb wires.
      The three principal methods of observing the azimuth of the plumb plane are as follows:
      (i) Exact complaining or alignment in which the theodolite is set up exactly in line with the plumb-bobs. This is a time-consuming procedure and requires much patience and manipulation.
      (ii) Weisbach Triangle method: In this method, the theodolite is set up in an approximate alignment with the plumb wires. A triangle is formed with the theodolite section and the two plumb wires, which is known as the Weisbach triangle. The small angle subtended at the theodolite station by the plumb wires, and the length of all the three sides of the triangle are measured very accurately. Measurement of other angles and distance to connect the three points with surface survey are also taken carefully. On the basis of these readings, the coordinates of the plumb wires are computed, which form the base for transferring the bearing to underground station.

(iii) Method of Weiss quadrilateral: A quadrilateral (known as Weiss quadrilateral) is formed by the plumb wires and the two stations in the shaft inset. Angles are measured from these two shaft stations to the plumb wires, and also the distance between the shaft station. From these observations it is possible to calculate all the polygonal angles and sides of the quadrilateral without any linear measurements, from which the bearing of plumb wires is computed and transmitted to another underground station.

(c) Precise magnetic correlation: The method involves the determination of the correct magnetic bearing of both surface and underground reference lines and the application of the difference between them to the azimuth of the surface line to give the azimuth of the underground line. A common coordinate is obtained by suspending a single wire in a shaft and running a traverse from it to the underground reference line. The method is prone to errors due to the presence of local magnetic disturbances, diurnal variations, magnetic storms, magnetic declination, instrumental peculiarities, etc.

(d) Gyroscopic Theodolite: Of late, a gyroscopic theodolite has become available. In this a gyroscope driven by an electric power-pack, whose axis is aligned to the true meridian forms the reference direction for the theodolite. The horizontal angles measured will then be able to give bearings directly.

5.5 Equipment and Norms of Surveying

The requirement of equipment for carrying out a survey for mineral exploration will depend upon the area to be covered, the time available for the work, topography of the area, the methods of surveying selected for use and the money available for buying the equipment. For carrying out triangulation, traversing, levelling, tachometry, differential GPS, plane tabling and linear surveys on the surface, the following equipment as given in Tables 5.5 and 5.6:

The average daily progress for surveying jobs will depend, apart from the nature of the terrain and the weather conditions, upon the skill of the surveyor and the persons assisting him and the type and condition of surveying equipment used. The above norms are, therefore, offered merely as a guide and may not represent the standard.

Table 5.5 Field Equipment for recommended for survey

S.L.	Name of instrument and description	Number of units required
1.	Theodolite, microptic with stadia diaphragm, telescopic metal tripod and optical plumbing device	1
2.	A pair of differential GPS with wireless connection	
3.	Engineer's tilting level with split bubble optical system	1
4.	Plane table approximately 65 cm × 65 cm with quick levelling head and telescopic metal tripod, a spirit level, compass and preferably a telescopic alidade	1
5.	(a) Steel tapes, graduated in metres and centimetres, of 30 m length (b) Metallic or linen tapes	1 2
6.	Levelling staves of telescopic or folding type graduated in metric system	2
7.	Trestles (wooden)	4
8.	Ranging rods	6
9.	Brunton compass with tripod	1
10.	Surveyor's umbrella	1
11.	Axe, wood chopper of bush knife, shovel, pick axe, hacksaw trowel, cement, galvanized 25 mm dia. pipe, etc., for clearing lines of sight and for levelling ground and marking permanent stations	
12.	Wooden pegs 50 mm × 500 mm for base line measurement and 25 mm for marking temporary stations	
13.	Hammer	
14.	Plumb-bobs and thread	
15.	Field books, pencils and eraser	
16.	Set of scales, protractor, set squares and drawing instruments	
17.	Torch light (3 cells)	

Table 5.6 Norms for various types of survey works listed below

	Description of work	Norm Average progress per day	Nature of terrain
1.	Boundary traverse with central and check-line which have been chained	330 m	Medium terrain
2.	Boundary levelling	1200 m	Medium terrain
3.	Boundary traverse by plane tabling	400 m	Medium terrain
4.	Detail filling and contouring by tacheometry, etc.	2 ha	Medium terrain

Medium Terrain = Terrain which is neither too flat nor too rugged

Further Readings

Breed GB, Hosmer GL (1959) The principles and practices of surveying, vol I. Elementary surveying. Wiley, New York

Kavanagh BF, Bird G (1996) Surveying—principles and applications. Prentice Hall, p. 257

Monteiro LS, Moore T, Hill C (2005) What is the accuracy of DGPS. J Navig 58:207–225

Punia BC (2017) Surveying and field work, vol 2. N.C. and R.K. Jain, Standard Book Depot, Delhi

Chapter 6
The Statistical Treatment of Exploration Data and Computer Application

6.1 Introduction

The analysis of exploration data essentially incorporates well-defined mathematical model describing the source, migration and accumulation of miner and trace elements in rocks. The use of statistics in exploration has increased, since the introduction of electronic computer in the early 1950s. Indeed, it is now often difficult to separate statistical methodology from its computational aspects. A geologist engaged in mineral exploration needs to be familiar with a basic knowledge of statistics. Here is a summary of statistical procedure, and how statistical methods are helpful to upgrade the interpretation of data gathered during the exploration campaign. There is no doubt that the statistical treatment of exploration data in the past decade has gained importance in the interpretation of data collected during an exploration campaign. Its importance can be understood by the increased difficulty of finding the "easy" deposit targets. More details of different methods are available as a list in further reading.

Rapid change and development have occurred in computerized data systems during the past few years. One significant change has been in the time-sharing computing and computer graphics. This gives an exploration geologist an access to the advanced computer programs for processing field data, and aid of computer graphics to display the results in a valuable form. Apart from computing, a major effort is in developing national geology/mineral inventory. It aims to provide for storage and exchange of all field data such as local and regional geologic data on map, collected over the past several years. Such data inventory system serves a link between organizers and individual exploration geologist, and may also attract more participation in exploration activity.

Our application and dependence on computer help in data handling and in exact calculation involved in the analysis of collected exploration data. Most of the processing and statistical treatment of exploration data can now be done on computer. To that extent, computer has made obsolete the calculation needed on the

G.S. Roonwal, *Mineral Exploration: Practical Application*,
Springer Geology, DOI 10.1007/978-981-10-5604-8_6

different statistical methods that have been enlisted in this chapter. The computer has thus made possible the processing of large amounts of data on a routine basis. It has helped to reduce the efforts in calculation and save time. But it needs to be mentioned that the successful application of computer-related procedures for data analysis in exploration needs cooperation of geologists and programmers. Here comes the role of geologist in understanding the various aspects of minor/trace elements distribution in a search area. This need has encouraged and successfully produced field data recording systems. They ensure the standardized recording of field observation and important data. This shall guide in locating mineralization, which is the aim of an exploration campaign. Statistical methods offer excellent grade computation procedures. Besides grade computations, these methods also help in studying the error involved in grade and tonnage computations. Since the parameters involved in these computations have wide applications in exploration, the subject will be dealt with at some length in the following paragraphs.

6.2 Statistical Methods and Applications

Procedures of statistical ore reserve calculation are given below:

Geological input	Exploration method, sample location, ore mineral composition/grade, dimension of the ore body
Ore reserve calculation	Geostatistical methods (application); Assay data frequency distribution, grade-tonnage for ore blocks, reserve category, grade level
Economics	Coefficients of extraction, cost, revenue
Reserve estimation and economics	cut-off grade, minable ore; reserve according to ore-blocks, metal percentage recoverable

The classical statistical methods do not take into account the degree of continuity of mineralization (sample-to-sample correlation in a statistical sense) and are concerned only with random, minimum trend data without any density contrasts. The methods commonly involve the use of normal or lognormal distributions. The steps involved in classical statistics are given in Table 6.1 and the important applications of classical statistical analyses are: (a) Computing the mean grade of ore; (b) Determination of the precision of grade estimate; (c) Determining the number of samples for specific precision levels of estimate, and computing the number of exploratory openings in mineral exploration; (d) Calculating the distance between two exploratory openings; (e) Calculating the number of exploratory openings; (f) Determining the sample volumes; (g) Regression and correlation and use of log normed distribution.

Table 6.1 Flowchart for statistical analysis

Assay Data

Screen for different assay population by minerals, grade of ore and geology

Assay frequency distributions for minerals and grade using unlocated assay value not considering actual xyz positions

Unimodal distributions indicating populations of assays	Multimodal distributions indicating subpopulations for other minerals or grade of ore
Compute statistics mean, variance, standard deviation, skewness	Screen for subpopulation oxide secondary, primary
If relative skewness is approximately 3, test for leg normal assay frequency distribution	Use log probability paper for cumulative assay frequency distribution plot to screen for sub population

If only a small number of assays are available use simulation techniques to generate larger assay populations to form better distributions

If skewness is present in assay frequency distribution try various transformations to normalize the distribution, square root, logarithms, binomial hypergeometric, etc.

Compare assay frequency distribution with theoretical mathematical distributions

Calculate confidence limits for normal and transformed normal assay frequency distributions

Match computed confidence limits with the acceptable limits specified as the goal for the sampling and evaluation

If confidence limits are acceptable for the exploration	If confidence limits are not acceptable and more work is needed
Calculate quadratic and cubic regression analysis to determine trends in grade thickness and mineralization	Use sample volumes—variance relationship $S_1^2\lambda_1 = S_2^2\lambda_2$ and standard error of the mean of calculating new sample volumes or new number of samples of same volume, and number of additional drill hole needed to match specified confidence limits calculate new drill-hole grid spacing as required by using standard error of the mean
Compute area of influence of ore assay	
Calculate statistical response surface and grade of ore interpolations for planning mining methods and production	
Deposit Evaluation	Phase II of Exploration

The statistical parameters to be determined are:

Variance: Variance is a measure of the dispersion of values x_i of the random variable (tenor values, assay values, assay x width values, thickness, etc.) about its mean $\bar{x}$. It is the expected value $(x_i - x)^2$.

$$\text{The variance} = \sigma^2 = \frac{1}{n}\left[\Sigma(x_i)^2 - \frac{\Sigma(x_i)^2}{n}\right]$$

The procedure for determining the variance is slightly different from the one shown in the earlier section on mathematical statistics.

Logarithmic variance: When the random variable shows lognormal distribution, the variance is found by converting the values to their natural logs. The logarithmic variance σ LN in such cases is expressed as follows:

Computing the mean grade of ore: as given under definitions
Determination of the precision of grade estimate

If the sample standard deviation is a good estimate of the population standard deviation, then the standard error of the mean $S_{\bar{x}}$ can be used to establish the precision estimate.

$$S_{\bar{x}} = \frac{\sigma}{\sqrt{n}}$$

where

$S_{\bar{x}}$ standard error of the mean,
σ standard deviation and
n number of assays.

Such estimates are expressed in terms of confidence interval at the 95% level (or other appropriate levels) of confidence, as shown below.

$$\bar{x} \pm t_{0.05} \times S_{\bar{x}},$$

where

$\bar{x}$ average grade of the ore (mean),
$t_{0.05}$ a table value for the sample size n at the relevant confidence level.

The confidence interval is a function of the standard deviation and the number of samples. The narrower the confidence interval, the better is the precision of the estimate.

6.3 Determining the Number If Sample of Specific Precision Levels of Estimate and Computing the Number of Exploratory Opening in Mineral Exploration

The confidence interval is a measure of the degree of confidence in the results of a sampling programme. When no data are available, a preliminary sampling programmes is undertaken to generate some data to establish various factors like $\bar{X}, \sigma, S^2, S_{\bar{X}}$, etc. Then the necessary sample volume for various confidence intervals can be calculated. This method can also be used for determining the number of boreholes/exploratory openings in an exploration programme. The method is shown in the following worked out example.

The sample chosen is from a bauxite exploration programme. The initial sample is of size $n = 61$, assays of alumina (Al_2O_3). The example is only illustrative (Table 6.2).

$$\bar{x} = \frac{f(\text{mp})}{n} = \frac{3060}{61} = 50.16\%\ Al_2O_3$$

$$\text{Sample variance} \quad S^2 = \frac{\sum t(\text{mp})^2}{n} - \left[\frac{\sum t(\text{mp})}{n}\right] = \frac{155{,}214}{61} - 50.16^2 = 28.47$$

$$\text{Standard deviation} \quad \sigma = \sqrt{S^2} = \sqrt{28.47} = 5.34$$

$$\text{Corrected (for population)} \quad \sigma = \sqrt{S^2} \times \frac{n}{n-1} = 5.34 \times \frac{61}{60} = 5.43$$

$$\text{Standard error of the mean} \quad S_{\bar{X}} = \frac{S}{\sqrt{n}} = \frac{5.34}{\sqrt{61}} = 0.68$$

Table 6.2 Assay data on frequency pattern

Al_2O_3 Assay data in frequency chart				
Al_2O_3 in grade interval of 3	Frequency F	Midpoint mp	f(mp)	$f(\text{mp})^2$
37.50–40.49	1	39	39	1521
40.50–43.49	6	42	252	10,584
43.50–46.49	11	45	495	22,275
46.50–49.49	10	48	480	23,040
49.50–52.49	10	51	510	26,010
52.50–55.49	12	54	648	34,992
55.50–58.49	8	57	456	25,992
58.50–61.49	3	60	180	10,800
	61		3060	155,214

$n = 61$

$$\frac{\text{CI}}{2} = \pm S_{\bar{X}} \times t_{0.05}$$
$$= 0.68 \times \text{table value of } t_{0.05} = \pm 1.34$$

With 95% confidence, the estimate of the average (mean) grade of the ore is 50.16 ± 1.34 or 48.82–51.50% Al_2O_3.

The value of confidence interval is 2.68 = 1.34 × 2. Here, it may be seen that the grade estimate is in terms of a range of values, rather than a single value of 50.16 which is called a point estimate. The point estimate of 50.16 implies an estimate of precision which is not attainable in grade computation because of grade fluctuations commonly seen in any group of assay values. The computation of a grade range with a known level of precision is one of the most important contributions of statistics to mineral exploration.

In this sample, instead of ±1.34, suppose a precision of say ±0.40 is desired. The number of samples (n) required for this can be determined by the formula.

$$S_{\bar{X}} \frac{\text{CI}/2}{t_{0.05}}$$
$$S_{1\bar{X}} = \frac{0.80}{2} / t_{0.05} = 0.20$$
$$\text{Now } n_1 = \frac{\sigma^2}{S_1 \bar{X}^2} = \frac{5.33}{(0.20)^2} = 711$$

Therefore, the total number of samples required for a precision of ±0.40 is 711.

If the original assay data had come from 61 boreholes, the total number of n samples computed above would have represented 711 boreholes. Similarly, the number of pits, trenches, etc., also can be computed.

In this method, a certain quantum of data is required before starting the computation. Such data may come from strategically placed boreholes, pits, trenches or any opening which can give a sample of 50–100 assay readings. Samples may also be chosen from blocks of ore which have already been proved/mined out.

The above formula does not have take into account the dimensions of the deposit. The formula given below incorporates the dimensions of the deposit also.

$$S_{\bar{X}} = \frac{\sigma}{\sqrt{(\text{APDH})(\text{NN})\left(\frac{\text{DW}}{\text{GS}} + 1\right)\left(\frac{\text{DL}}{\text{GS}} + 1\right)}}$$

where

$S_{\bar{X}}$ standard error of the mean of a sample,

σ standard deviation of the sample consisting of assays above the cut-off,

APDH average assay per drill hole,

NN ratio of the number of assays above cut-off grade, to total number of assays, i.e.

$$\frac{\text{Number of assays above cut off}}{\text{Total number of assays}}$$

DW width of the deposit,
DL length of the deposit, and
GS grid spacing.

Here, instead of the n number of holes, the actual distance between borehole or grid spacing is calculated.

There are various other methods and procedures which can be used for determining the number of exploratory openings or the grid spacing, which make use of parameters like variance, coefficient of variability, etc. Some of these formulae are described below.

6.3.1 Calculation of the Distance Between Two Exploratory Openings (Boreholes/Pits/Trench)

$$\perp = \frac{p^2}{W^2} \times L \quad \text{or} \quad \frac{p^2}{W^2} \times A,$$

where

$\perp$ the average distance between two boreholes,
p accuracy of the relative mean error in the percentage of the arithmetic average,
W variance of the deposit calculated by the formula

$$W = \sqrt{W_{\mathrm{m}}^2 + W_{\mathrm{c}}^2 + W_{\mathrm{d}}^2}$$

where

W_{m} variance with regard to the thickness of the ore body,
W_{c} variance with regard to the assay value,
W_{d} variance with regard to the bulk weight,
L length of the sector/bench/block/deposit to be explored and
A are of the sector/bench/block/deposit to be explored.

Here, the main problem is to calculate the value of 'p' from a known deposit or a block or ore, and then to substitute this value for various similar deposits/blocks of ore. In a wholly unknown block, in order to determine the value of 'W' certain strategically placed boreholes/trenches/pits would be necessary in the beginning.

6.3.2 Calculating the Number of Exploratory Openings

Formula 1

The number of boreholes/pits/trenches required can be computed by using the following formula:

$$N = \frac{W^2}{P^2},$$

where

W and P	are as described above and
N	number of boreholes/pits/trenches, etc.

Formula 2

Here N is calculated on a different basis:

$$N = \left(\frac{Q}{a}\right)^2 \times P,$$

where

N number of boreholes/pits/trenches required for a specific precision of estimate,
a precision of estimate,
P probability factor which is 2 for normal distribution and
Q Coefficient of complication of the deposit determined by the formula

$$Q = \frac{V}{K \times M}$$

where

V coefficient of variation of the thickness of the deposit (coefficient of variability described earlier),
K coefficient of variation of impurities in the deposit determined by the formula

$$K = \mathrm{OC} - (F_1, F_2, F_n \text{ etc.})$$

where

OC	total surface area of the deposit,
F_1, F_2, etc.	surface areas of impurities within the deposit,
M	modulus of complication of the ore waste contact determined by the formula

$$M = \frac{\text{PE}}{\text{PC}},$$

where

PE perimeter for theoretical ellipose of the deposit outline, determined by the formula

$$\text{PE} = 2\pi\sqrt{1/2(a^2 + b^2)}$$

where

'a' and 'b' are the maximum length and width of the deposit, respectively and
PC circumstance of the deposit contact with waste (actual perimeter as measured).

For computing this formula, an initial sample is required which may be obtained from an initial set of strategically placed boreholes. Data from a known similar deposit can also be used.

6.3.3 *Calculation of Sample Value*

Statistical methods can be used for predicting the desired sample volume in producing a pre-required sampling precision. The relationship between the sample volume and variance is

$$\text{The formula is } S^2\lambda^1 = S^2\lambda^2, \quad \text{where } \lambda^1 = \text{sample volume}$$

Since S^2 (variance) is related to the standard deviation $\sigma = \sqrt{S^2}$ and to the standard error of the mean and confidence interval $\left(S_{\bar{X}} = \frac{\sigma}{\sqrt{n}} \text{ and } \frac{\text{C.I.}}{2} = S_{\bar{X}} \times t_{0.05}\right)$, we can calculate the values of λ (sample volume).

For various values of confidence interval, new values of $S_{\bar{X}}$, σ and finally S^2 are determined and the value of S is substituted in the relationship $S_1^2\,\lambda_1 = S_2^2\,\lambda_2$ to arrive at new sample volumes.

6.3.4 *Regression and Correlation*

By studying the relationship between the various variable, it is easy to predict in a mineral deposit new information from a set of exploration data. The techniques used are regression and correlation analyses. The variables whose interrelationship is to be studied may be grade or whose interrelationship is to be studied may be grade of assay values or specific gravity, thickness, width, mineralogy of the ore, micro or macrostructure, etc. Any variable/variables showing predictable relationship can be used to generate reliable new data. The variables may be independent or interdependent. Various regression techniques which are needed in such cases are dealt with below.

Suppose the relationship between the grade (x) and bulk density (y) of an iron ore deposit is being studied.

After plotting the information, the relationship has been found to be in the form of equation.

$$y-(a+bx) = O,$$

where

x grade of ore,
y bulk density,
a and b are the coefficients of linear regression.

For purposes of prediction, it is necessary to find the values of 'a' and 'b' using the following formulae:

$$a = \frac{\sum X^2 \sum y - \sum x \sum xy}{n\sum x^2 - [\sum x]^2}$$

where n is the number of pairs of observations, and

$$b = \frac{\sum xy - \sum x \sum y}{n\sum x^2 - [\sum x]^2}$$

The equation of the line of regression of y and x is

$$y = \left[\frac{\sum x^2 \sum y - \sum x \sum xy}{n\sum x^2 - [\sum (x)]^2}\right] + \left[\frac{\sum xy - \sum x \sum y}{n\sum x^2 - [\sum x]^2}\right] x$$

After establishing this formula, new values of 'y' such as $y_1, y_2, \ldots, y_k$ etc. can be computed by knowing only the value of $x_1, x_2, \ldots, x_k$ etc.

In the case of mineral deposits, the relationship between various variables is seldom independent. When various variables show a correlation, the formulae used are more complex. A typical equation showing the relationship between 'y' dependent variable to various independent variables, like $x_1, x_2, x_3, \ldots, x_k$ is given:

$$y = bo + b_1x_1 + b_2x_2 + \cdots b_nx_k$$

where

b_0 is a constant and
$b_1, b_2, \ldots, b_k$ are partial regression coefficients.

The values of $b_0, b_1, b_2, \ldots, b_k$ are found by solving the basic equations. The number of normal equations increases with the number of coefficients in the type equation.

The basic equations are as follows:

$$\Sigma y = nb_0 + b_1\Sigma x_1 + b_2\Sigma x_2$$
$$\Sigma x_1 y = bo\Sigma x_1 + b_1\Sigma x_1^2 + b_2\Sigma x_1 x_2$$
$$\Sigma x_2 y = bo\Sigma x_2 + b_1\Sigma x_1 x_2 + b_2\Sigma x_2^2$$

Such equations are most amenable to standard computer treatment.

After finding out the numerical values of $b_0, b_1, b_2, \ldots, b_k$ etc., it is possible to generate the values of $y_1, y_2, y_3, \ldots, y_n$ on the basis of various sets of $x_1, x_2, x_3, \ldots, x_k$ values. These kx's are n-vectors.

This can be applied in the following way. Let 'y' bet the gibbsite content of a bauxite sample. 'y' shows dependent four variables x_1 = Al_2O_3 content, x_2 = L.O.I., x_3 = Fe_2O_3, and x_4 = TiO_2 + minor minerals content, Gibbsite = $y = b_0 + b_1$ Al_2O_3 + b_2 LOI + b_3 Fe_2O_3 + b_4 TiO_2.

By solving the basic equation for b_0, b_1, b_2, b_3 and b_4, the gibbsite content can be predicted without costly and time-taking X-ray analysis which is normally required.

When the data are in three dimensions, as is often the case with mineral exploration data, a procedure known as the 'Three Dimensional Regression Analysis' is resorted to. Here, the Cartesian (graph) coordinate values (values of X, Y and Z Cartesian coordinates) are introduced into the regression equation.

The equation may take the following form:

$$\bar{x} = a_1 + a_2 x + a_3 y + a_4 z$$

where

$\bar{x}$ mean value at any point, x, y, z.
$a_1, a_2, \ldots$ etc., regression coefficients the values of which are a function of the strike, dip and plunge of the linear surface.

The equation is re-written for ease in mathematical treatment, as follows:

$$X = a_1 + a_2(x - \bar{x}) + a_3(y - \bar{y}) + a_4(z - \bar{z}) + E$$

where

E error term associated with any value of X.

This formula is used in a variety of forms in mineral exploration.

6.3.5 Coefficient of Correlation

In all the above mathematical considerations, the underlying principle is the degree of interrelationship between the variables. This relationship can be expressed by a mathematical expression, the correlation coefficient. The formula for this is as follows:

$$r = \frac{n\Sigma xy - \Sigma x\Sigma y}{\sqrt{\left[n\Sigma x^2 - (\Sigma x)^2\right]\left[n\Sigma y^2 - (\Sigma y)^2\right]}}$$

where

X and Y	are variables,
n	is the number of observations, and
r	is the linear correlation coefficient.

The coefficient of correlation r may have positive or negative values, ranging from −1 to 1. When no correlation exists, the 'r' value may come close to zero. As the degree of correlation improves, the 'r' values may approach one. When the value is positive, the variables vary directly. If the value is negative, the variation is in inverse proportion.

These methods are extremely useful when studying core losses in drilling and choosing regional exploration targets.

6.3.6 Use of Lognormal Distributions

It was mentioned earlier that some data on being arranged in a frequency distribution show skewness. The degree of skewness can be tested by a formula.

$$\text{Coefficient of skewness} = \frac{\text{mean} - \text{mode}}{\text{standard deviation}}$$

When the numerical value of the coefficient of skewness is less than 0.3, the distribution is lognormal. The condition of lognormality can be tested by other methods also like plotting the histogram and plotting cumulative frequency (if the cumulative frequencies are plotted on a log probability paper, the logtransformeddata will form a straight line).

In the latter case, there are few points to be observed:

(a) A change of slope of the line indicates a bimodal distribution.
(b) The accuracy at the top end of the curve is usually poor, because accuracy of assaying high values is poor.

(c) This plot can be utilized to determine sample parameters such as geometric mean, variance and confidence limit.

The skewed distributions can be attempted to be converted to normal distribution by transforming the assay readings to their nature logarithmic value. After this conversion, if it conforms to a normal distribution, all equations applicable to normal distribution hold.

However, in the case of lognormal distribution, the arithmetic mean is a poor indicator. In its place, the geometric mean need be determined and the standard deviation, standard error of the mean worked out.

In the case of lognormal distribution, the best mean grade estimator is a quantity known as the Sichel's '*t*' estimator.

This is expressed as follows:

$$t = \mathrm{e}^{\bar{x}} x(v),$$

where

$\bar{x}$ mean of the natural logarithms of the assays,
v natural variance of the natural logarithms of assay, and
$r(v)$ a complex mathematical function.

6.4 Geostatistical Methods and Their Applications

Geostatistical methods are also utilized in grade estimation. The basic tool in geostatistical studies is the variogram. In this, the spatial correlation between the various sample values is studied by means of a variogram is a mathematical function defining the natural (intrinsic) dispersion of assay values as shown by log arithmetic variances.

$$\sigma^2 \mathrm{LN} = \frac{5.302}{n}\left[\sum_{i=1}^{n} fi(\log xi)^2 - \left(\frac{\sum_{i=1}^{n} fi \log xi}{n}\right)^2\right]$$

Standard deviation: The standard deviation is the square root of the variance and is expressed by 'σ'.

Standard error of the mean: The standard error of the mean is defined by the formula as shown in an earlier section

$$\sigma(\bar{x}) = \frac{\sigma}{\sqrt{n}}$$

These parameters can be used for determining the error involved in various types of grade, tenor and tonnage estimates by analyzing the relevant exploratory data. The procedure for making the estimates are described below.

As mentioned earlier, the dispersion laws of various deposits are different. Two very common types of dispersion laws are represented by the De Wijsian model and the Transitive model.

6.4.1 The De Wijsian Model

The general De Wijsian law is stated as follows:

$$\sigma^2 \mathrm{LN}(t) = \alpha \,\mathrm{Ln}\left(\frac{V}{v}\right)$$

where

$\sigma^2 \mathrm{LN}(t)$	logarithmic variance of tenors,
α	a geostatistical factor of dispersion of mineralization,
V	volume of the ore block, and
v	volume of the samples taken in this block the shape of which is the same as that of the block.

The quantity α can be determined by three methods.

(1) From the slope of the variogram.
(2) From Matheron–De Wijs formula given below:

$$\sigma^2 \mathrm{LN}(t) = 3\alpha \,\mathrm{Ln}\left(\frac{D}{d}\right),$$

where

$\sigma^2 \mathrm{LN}(t)$	logarithmic variance of the samples (obtained from statistics),
α	absolute coefficient of dispersion,
D	linear equivalent of the block of ore considered,
$D = A + B + 0.7\,C$	where A, B and C are the three dimensions of the ore block, and
$D = a + b + 0.7\,c$	where a, b and c are the dimensions of the diamond drill hole, (used for computing $\sigma^2 \mathrm{LN}(t)$).

(3) From the variance—variogram chart prepared by Matheron. Of these, the Matheron De Wijs formula described above appears most useful for practical purposes.

The formulae described above can be used for calculating the total error involved in any ore estimate based on the exploration data. The calculation is done on the principle of "Extension geovariance".

6.4.2 Extension Geovariance

It is the normal procedure to extrapolate the assay value of a diamond drill hole, channel sample, etc., to its natural zone of influence. However, unless the mineralization is very uniform, an error is introduced in such extra extrapolation. This error is related to three factors.

(i) Variability of mineralization,
(ii) Geometry of the sample and
(iii) Zone of influence of the sample,

Extension geovariance measures the error involved in this extrapolation quantitatively taking into account the geovariance has three error components, which are as follows:

(a) Line geovariance: Error committed when the values of a series of points are extended to a line, e.g. the assays of a linear series of grab samples on an ore face.
(b) Section geovariance: Error committed when extending the value of a line to its rectangle of influence, e.g. weighted average grade of a drill hole intersection.
(c) Block geovariance: Error committed when extending the average grade of a section to its volume of influence, e.g. the weighted average grade of a series of drill holes in a section.

In addition, the sampling variance which indicates the error committed in various sampling practices also contributes to the total error in the estimate. For finding the extension geovariance, the formula is

$$\sigma\text{LN}(E) = \sqrt{\frac{\sigma^2(s)}{\bar{x}^2} + \sigma^2\text{LN}(L) + \sigma^2\text{LN}(x) + \sigma^2\text{LN}(B)}$$

where

$\sigma\text{LN}(E)$ extension geovariance (total error in the estimation),
$\frac{\sigma^2(S)}{\bar{x}^2}$ sampling variance,
$\sigma^2\text{LN}(L)$ line geovariance,
$\sigma^2\text{LN}(X)$ section geovariance, and
$\sigma^2\text{LN}(B)$ block geovariances.

Since grab sampling is seldom used, $LN^2(L)$ does not usually appear in the computation.

$\frac{\sigma^2(S)}{\bar{x}^2}$ is determined by dividing the variance by square of the sample mean $\bar{x}$.
$\sigma^2 LN(X)$ is found by using the formula:

$$\sigma^2 LN(X) = \frac{\pi}{2}\alpha\frac{h}{p}.\frac{1}{N}$$

where

N is the number of drill holes intersecting the relevant section,
h interval between the drill holes,
p average length of ore intersection, and
α absolute coefficient of dispersion.
$LN(B)$ is found by using the following formula:

$$\sigma^2 LN(B) = \frac{\pi}{2}\alpha\frac{H}{L^1}.\frac{1}{N^1} = \frac{\pi}{2}\alpha\frac{H}{L}.\frac{L}{L} = \frac{\pi}{2}\alpha\frac{S}{L^2},$$

where

N^1 number of sections—say levels in a block of ore,
L^1 average strike length of the deposit in horizontal section,
L total length of ore recognized in various levels,
H level interval and
S surface of the ore body in a longitudinal section.

6.4.3 The Transitive Model

The intrinsic dispersion conforms to the following law:

$$\Upsilon(r) = \text{Co} + \text{CT}(r, a)$$

where

Υ distance,
a range of correlations,
Co nugget effect, i.e. value of $\gamma(r)$ at $\gamma = o$,
C a constant equal to the difference between the variance of assay values and the nugget effect, and

$T(r, a)$ pure transition function which is equal to r/a for $r \leq C$ and equal to one when $r > a$.

In practical terms, this law describes the geometrical fragmentation of various grades of mineralization in a deposit. When two samples M and M' are close to each other, they are likely to be in the same grade range (grade micro-basins). When the distance r between these two increases, the two samples are likely to be in different grade ranges (grade micro-basins). This formula can also be used for calculating the extension geovariance (total error of estimation).

For calculation of the geovariance, the following sequence is adopted. In an estimate of an ore block, the value of length '*f*' of intersection is extended to its rectangle of influence o of width '*h*' the intrinsic law $T(r, a)$ is replaced by $F(h, f)$ which is an auxiliary function representing the geometrical equivalent of the block.

The earlier formula now becomes

$$\sigma^2(t) = \mathrm{Co} + \mathrm{CF}(h, L),$$

where

$\sigma^2(t)$ relative arithmetic variance of assay values (obtained from statistics),
Co nugget effect,
C pitch of the transitive variogram, and
$F(h, L)$ value from charts for h/a and $1/a$, '*a*' being the range of correlation of values as shown by the transitive variogram:

$$\text{The total/error of estimate} = \sqrt{\text{Section geovariance} + \text{block geovariance}}$$

Section geovariance is determined by the following formula:

$$\text{Section geo-variance} = \frac{1}{\mathrm{NP}}\sigma^2(e)\left(\frac{L'}{N}\right),$$

where

N average number of ore intersections in each level,
L' average length of ore in each level, and
P number of levels intersected by diamond drill holes.

$\sigma^2(e)h$ which is an elementary estimation of variance of a segment of length '*h*' and assumes three values depending on '*h*' the length of the segment for which the estimation is made.

Thus, $\sigma^2(e)h$ when $h < a = \frac{1}{6} - \frac{h}{a}$
when $a < h < 2a = \frac{1}{2}.\frac{h}{a} + \frac{a}{h} - 1 - \frac{1}{3}\frac{a^2}{h^2}$
when $h > 2a = 1 - \frac{a}{h} - \frac{1}{3}\frac{a^2}{h^2}$ and

value of $F(h) = \frac{1}{3}\frac{h}{a}$, when $h < a$ and
$1 - \frac{a}{h} + \frac{1}{3}\frac{a^2}{h^2}$, when $h > a$
Block geovariance = $\frac{1}{p}\sigma^2(e)$ (H, L)

where H is the standard vertical interval between the levels. When H is smaller than a (range) then

$$\sigma^2(a)(H,L) = 0.07385\frac{h^2}{aL} - \frac{h^2}{40\,aL^2}$$

$\sigma^2(e)$ (H, L) is the value of the function when $H > a$.

6.4.4 Krigging and Its Application in Grade Estimation

In certain special cases, it is difficult to make a reliable estimate of the grade of ore in small blocks in working mines due to absence or inadequacy of data within the block. In such cases, a procedure known as continuous Krigging is needed for estimating the grade.

The case dealt with here is that of a block of ore which has little data of its own, but is surrounded by levels and drives which have data. The usual procedure in such cases is to find out the weighted panel averages around the block and interpolate the results. To increase the efficiency, the following procedure (continuous Krigging) is adopted.

The ore block (square) is assigned additional weighting factors in such a way that the weight for the square block does not exceed 1000 (250 for each side). Grade computation from the surrounding blocks is now given an additional weighting factor, the nearest factor of each reading. Now the weighted mean is found out. This yields the optimum weighted average of the square block of ore.

Another procedure known as discontinuous Krigging is used, when the block of ore in question has been explored by boreholes put in square grid. The problem is to ascertain the influence of each borehole.

The central borehole has an average grade of u, the first ring of boreholes has an average weighted grade of v and the outermost ring of boreholes an average weighted grade of w.

Here, an estimator $z = 4$ is made use of to predict the grade of the central block. The estimator z is known as the Krigged estimator and is considered to be the most efficient under the circumstances.

The formula for Krigg estimator is

$$z = (1 - \lambda - \mu)u + \lambda v + \mu v$$

here

z Krigg estimator of the panel considered,
u weighted average grade of the central hole,
v weighted average grade of the first ring of holes,
w weighted average grade of the second ring of holes, and
λ and μ geometric coefficient the values of which are read off from a standard chart,

$$x = \frac{h}{a} = \frac{\text{average thickness of ore (holes } u, v, \text{ and } w)}{\text{cell edge of basic square drilleing grid}}$$

The chart is prepared by putting the values of $x = \frac{h}{a}$ on the 'x' axis and the values $\frac{1}{3\alpha}\sigma^2 k$ on the 'y' axis.

Here α = value of absolute coefficient of dispersion determined from the variogram, and $\sigma^2 k$ = variance of each intersection plotted on 'x' axis. The corresponding value of $\frac{h}{a}$.

Geostatistical methods described above are of use in certain types of mineral deposits like gold, copper, lead–zinc, etc. The use of this method for other mineral deposits is under experimentation. The methods find the best application in producing mines where grade control has to be done continuously. However, if we deal with issues of variability, krigging has a downside. The means of achieving minimum estimation variance in krigging is to smooth the values. This is a consequence of the information effect and implies that the estimated block values have a lower variance than the true block values.

Such smoothing is necessary to minimize conditional bias. Geostatistical simulation methods preserve the variance observed in the data, instead of just the mean value, as in interpolation. Their stochastic approach allows calculation of many equally probable solutions (realizations), which can be post-processed to quantify and assess uncertainty. Many practitioners are suspicious of stochastic methods—and even reject them outright—because natural processes that form reservoirs are not random.

Further Readings

Clark I (1982) Practical geostatistics. Applied Science Publishers, London, p 129
Davis JC (1973) Statistics and data analysis in geology. New York, Willey, p 550
David M (1977) Geostatistical ore reserve estimation. In: Developments in geo-mathematics, vol 6. Elsevier Scientific Publishing Company, Amsterdam, p 364
David JM (1988) Handbook of applied advanced geostatistical ore reserve estimation. In: Developments in gee-mathematic, vol 6. Elsevier Scientific Publishing Company, New York, p 232
Deutsch CV, Journel AG (1992) GSLIB: geostatistical software library and user's guide. Oxford University Press, NY
Goovaerts P (1997) Geostatistics for natural resources evaluation. Oxford University Press, pp 284–286

http://petrowiki.org/GeostatisticaL_conditional_simulation
Isaak HE, Srivastava RM (1989) An introduction to applied geostatistics. Oxford University Press Inc., p 559
Koch GS, Link RF (1986) Statistical analysis of geological data, 2nd edn. Wiley, New York, p 375
Sahu BK (2005) Statistical models in earth sciences. B, S. Publication Hyderabad, India, p 211
Vann J, Guival GD (1998) Beyond ordinary Kriging: non linear geostatlstical methods in practice. In: Proceedings of symposium on Oct, 30, 1998; Beyond ordinary Kriging, An overview of non-linear estimation
Wellmer FW (2012) Statistical evaluation in exploration for mineral deposits. Springer, Berlin, p 375

Chapter 7
Interpretation of Exploration Data

7.1 Interpretation and Evaluation

Interpretation and evaluation aims at an assessment of the progress in exploration, and the determination of the commercial value of a mineral deposit. In both, the methodology is similar although the precision of evaluation needs to be of a higher order in the latter case. Exploration is carried out in various stages thus it is necessary to check periodically the progress at all steps whereby the deposits which are not of immediate economic value can be relinquished. Gaps seen in exploration need to be identified to resolve systematically. In this way, future planning is dependent on evaluation. However, the final evaluation of a deposit is made at the stage of taking it up as a commercial prospect for mining. Evaluation of a deposit for this purpose is made in two stages: (i) Interpretation of data, and (ii) Computation and classification of reserves and grades.

Exploratory data need to be evaluated continuously as operations progress on until the campaign is completed. The former is done step-by-step because the data interpretation is done after the data are assembled. These operations and processing of data generated in an exploration operation are

(a) Survey data, (b) Surface and subsurface geological data, (c) Surface and subsurface drilling data, (d) Trench and pit data and (e) Exploratory mining data.

For purposeful interpretation, plans and sections are necessary. The important amongst them are structure contour plan, isopach (isochore) plan, isograd plan, assay plan, ore distribution plan, cross section of ore bodies, isometric projection and slice plan. These are explained in the following paragraphs.

G.S. Roonwal, *Mineral Exploration: Practical Application*,
Springer Geology, DOI 10.1007/978-981-10-5604-8_7

7.1.1 Structure Contour Plans

Structure contour plans are of two types: (i) floor contour plans and (ii) roof contour plans. They depict the configuration of the floor or roof surfaces by contour lines by joining points of equal elevation. They guide in interpreting the structure of sedimentary rocks. Often a key ore bed is chosen for drawing the contours. The principle involved in constructing a plan is similar to surface contouring, the difference being the points of elevation of a particular horizon (roof or floor) to be computed from exploratory data. Structural contour plans would help in interpretation, in reserves estimation, more so in open pit mining.

7.1.2 Isopach (Isochore) Map

Isopach maps exhibit variation in the defined thickness of a formation by contour lines drawn through points at which they are of equal thickness. Isopachs can be prepared for deposits of any shape when sufficient data are available. An isochore map is an isopach map drawn on the drilled thickness of the strata without referring to its true thickness. It is drawn for formations with gentle dips. The principle is similar to preparing a structural contour map but actual contours are drawn on thickness rather than a reference horizon. An isopach map alone or in combination with other plans have interpretative value in estimating depth-wise reserves, and in planning open pit mines.

7.1.3 Isograd Plan

These plans depict the grade quality of a mineral deposit. Like the isopach plan, it is prepared for ore bodies of shape. The principle is similar to preparing isopach plan. Here, the thickness values are substituted by grade, individual metal value and combined metal quality values. Similar to isopach, this helps in interpretation and is used in estimating gradewise reserves. It can be used in combination with isopach or floor contour or with slice plan and face plan.

7.1.4 Assay Plans

Assay plan shows the results of sampling, useful for underground openings with full assay detail. This plan needs to be prepared on large scales (1:500 or even larger, 1:200 is the most common in vein type deposits). Along with the assay plan, there is need for a plan on similar scale showing the geological and structural details of the area, the positions of boreholes and other exploratory openings. Such an assay plan helps in delineating the ore body based on different assay cut-offs, drawing up mining plans, quality and grade controlled production program. This is an important plan which helps in evaluation.

7.1.5 Ore Distribution Plan

An ore distribution composite plan shows mining face. Mining data such as position of benches, faces in addition to the exploration details. Different grades and types of ores are shown together with their thickness and tonnages. Plans are formed on a slice projection, bench plan in the case of opencast mines, or in the case of face and stope plans in the case of underground mines.

7.1.6 Cross Section

For the purpose of constructing cross sections of an ore body, the principles involved are akin to geology plans. Here higher precision is needed, and thus more data are required. Generally, cross sections along the dip of the ore body are called traverse sections. Cross sections along the strike are called longitudinal vertical sections. Transverse sections are used for interpretation. They show the underground structure and ore distribution of the deposit. By projecting bore hole data the quality and grade information can be marked on cross sections. The depth of mineralization can be shown graphically where bench position projected, open pit mines and economic quarry limit decided. Cross sections help in computing the reserves. One difficulty in construction of cross sections is determination of correct dip along the section line. When line of section is at right angles to line of strike, true dips is used. Generally, line of section makes an angle with the line of strike, the section can be constructed with the apparent dip component along line of the section. Here, it is necessary to convert the true dip into the apparent dip.

7.1.7 Longitudinal Vertical Projection

Longitudinal vertical projections are drawn along the strike of the ore body, selecting a common surface reference point along the strike. According to the purpose for which the sections are chosen line may pass along the footwall, hanging wall contact or the centre of the ore body. All boreholes, pits trenches and subsurface details are brought on to the longitudinal vertical section by suitable projections to enable details to be viewed in one plane. Together with the transverse sections, it helps in depicting the ore body in three dimensions, which helps understand the disposition of the ore body.

7.1.8 Isometric Projection

Isometric projections guide to depict structure and disposition of complex ore bodies. Such projections give three-dimensional view in display of the ore body underground. The isometric projections are made on a cube which is also represented by a single projection. This brings out a perspective of an ore body as seen from one angle where only three sides of figure can be seen instead of all the six sides. Such projections are drawn after data from the surface and from underground are plotted on a geological plan.

7.1.9 Slice Plan

Here, plans show an outline of different slices in an open pit mine, indicated in each slice geological and mining details. The geological details include structural features, ore types, quantity, determined during exploration and face sampling. It shows the position of benches as they advance, ore and waste and other information on a plan. Here, the height of each slice equivalent to the height of the proposed benches in mining is kept.

Apart from plans and sections, there are geophysical and geochemical maps if such surveys have been carried out. Every section and projection help in correlation and interpretation of the exploratory data. Since all the exploratory data cannot be depicted in a small plan, it is recommended to subdivide the plans. Such plans can be made on a copy of the base map of common grids for correlation and can be traced on tracing cloth so that one can be superimposed on other. Even after detailed sampling and exploration, some aspects of ore body may be needed. For reliable on shape, size, grade and tonnage of the ore body, it is important to fill such gaps. It can be done by correlation and interpretation. There may be two points where outcrops are clearly visible but, in between, the outcrop is not seen. By interpreting the strike, dip and structure of the ore body, the missing outcrop can be placed on the map by interpolation. This principle is adopted in various situations where specific data gaps need to be bridged up. By representing the ore body in three dimensions, a type of interpretative correlation is established.

7.2 Computation and Quantitative Assessment of Mineral Potential

Here, examine determination of ore grade and tonnage. In the above section, different aspects have been discussed. We now examine ore reserve estimation. The ore reserve estimation is based on ore grades and tonnages of the mineral prospect as determined during the exploration process. A final sampling and data generated would demand evaluation. The reserve estimates take into account the geological controls of the mineral as it occurs and shows its pattern of distribution. In the final analysis, it is not only reserve calculation but tonnage, cut-off and metal content in relation to economy—and investment—return basis. Therefore, final synthesis shall examine influence of mining and metallurgical controls. This is accuracy and reliability of estimates and a successful processing the ore and its reconvertibility into a successful venture comes in. Finally, it is the price trend and financing, mining, processing cost, that would govern whether the existing ore is graded and enough to make this a successful and profitable mining operation.

The metals of ore reserve estimation depend on a variety of factors, such as (a) the nature of the ore itself/and what metals can be extracted, (b), the nature of the mineral deposit, (c) the configuration in such as geometry, homogeneity of ore and its distribution in the host rock.

All these help reveal the size of the deposit and nature of ore—which would guide the type of sampling spacing needed for fully reliable results on estimates of reserve quantity.

There would be a certain degree of deviation from calculated averages but as long as they are within acceptable range, it is fine. Assaying of ore quality and grade would help in perfect calculation of grade and tonnage of the ore deposit. To arrive, at this complete evaluation of the ore grade, its dimension and quantity—will govern the production and life of mine. Production will primarily be governed by economic consideration. Three options available to achieve this are: (a) Geometric method, (b) Statistical methods and (c) Computer application as have been discussed in Chap. 6.

The Geometric Method of Reserve Computation

This depends on (i) the volume of ore that can be calculated by geometric models. This shall be dependent on distribution pattern and stage of evaluation, initial or expansion; (ii) the density of the ore and finally (iii) the limitation/difficulty anticipated during mining. This is best achieved by adhering to estimate correction factor.

$$Q = V \times d \times l_0,$$

where

V is volume in m^3
d density of the ore in t/m^3
l_0 correction factor estimated—to look for possible losses during mining the metal content (M) in metric tonnes (t) is

$$M = Q \times Z \times l_m,$$

where

Q ore reserves
Z is the ore grade in % (can be expended upto decimal level—such as 30 or 0.30%

l_m is an estimate correction factor to account for the losses of metal recover during processes (cost: $0 < l_m < 1$).

In geometric reserve calculation, each sample or drill intersection of the ore barely is assigned an area and volume of influence by makings geometric models—rectangle, triangle, polygon or profiles, and then their three dimensioned equivalents as shown in Fig. 7.1.

Such volume and assay value are taken to work out ore grades and tonnages of each block. One assumption in geometric estimate is that the area of influence is dependent on the distance with respect to the previous sample location. Such assumptions have limitation especially when the ore occurrence is not homogenous. This progressive close grid pattern is often adopted to overcome difficulty of best estimates. Because of the subjectivity and difficulty in exact estimation, are superseded by application of statistical and computer simulation methods. In spite of such difficulty, this method is still in practice for a preliminary assessment. The methods are in use partially in small-scale mining operation.

Once a deposit has been proved by exploration, it becomes more or less a commercial proposition. The deposit is now viewed in terms of investments, return on capital, daily rate of production. The deposit has to support a mine to yield ore in enough quantity and grades to make it profitable within a fixed period of time at a rate of production to yield a profit acceptable to the investor. In order to take investment decisions, an accurate estimate of theore reserves together with their grade is essential. All exploration activities lead ultimately to such an estimate. The more precise the exploration the more accurate the reserve estimate, the reserve can be estimated by computing the volume of the ore and converting that volume into tonnage as discussed earlier.

$$\text{Tonnage} = \text{Volume} \times \text{Sp.Gravity(Bulk Density)}$$

i.e.

Q V D
Q tonnage
V volume of the ore body and
D specific gravity or density of raw mineral materials (bulk density)

The computation of the volume of the deposit is, however, complicated by factors like geology, size and shape of the deposit and pattern of mineralization which are inherent in all deposits. Within these broad limitations, a large number of methods are available for the computation of volume. However, some assumptions regarding the behaviour of the ore body and the nature of exploratory data are necessary for any reserve computation. They are:

(1) Any basic character of an ore body, established at one point by exploration, changes or extends to an adjoining point in accordance with certain principle such as the rule of gradual change or the law of linear function. According to this rule, all parameters (such as thickness and grade) of an ore body expressed

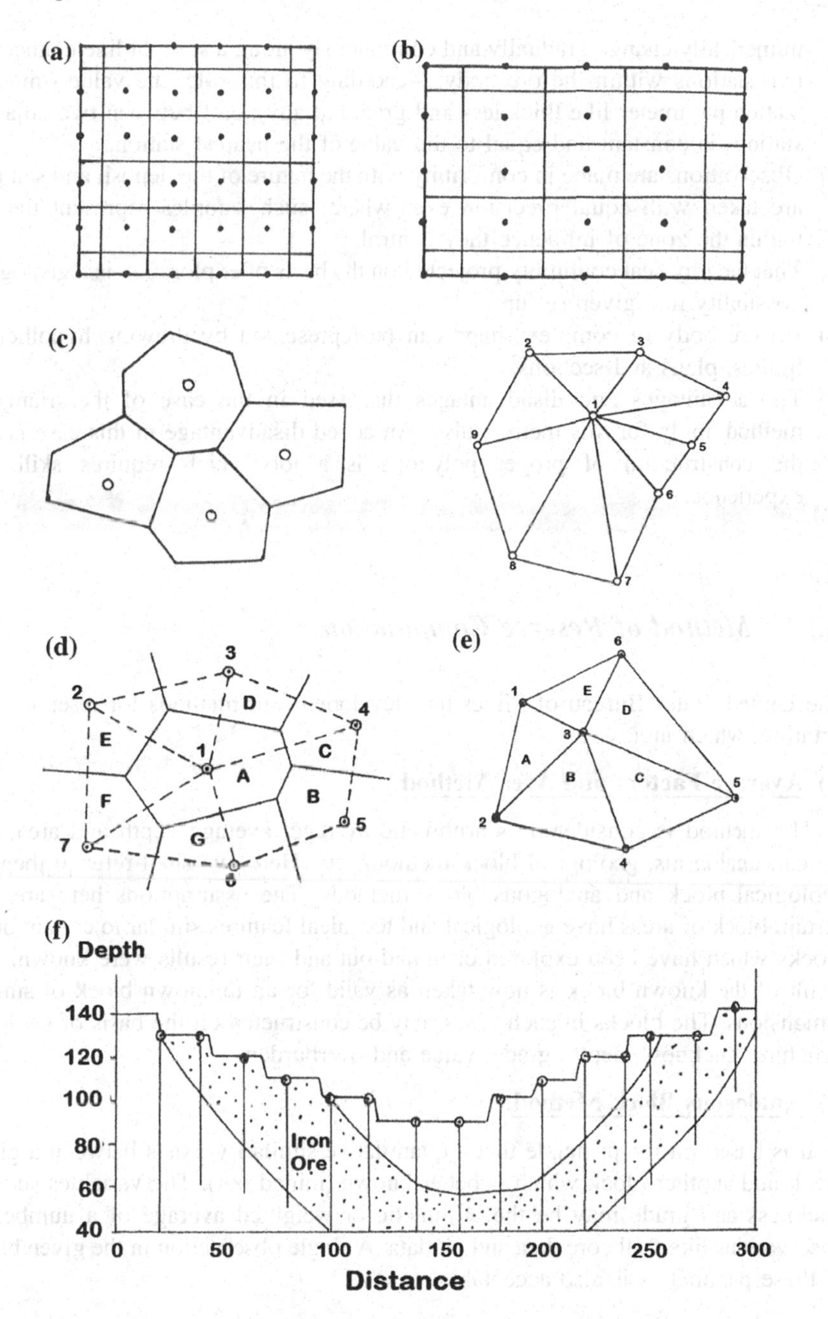

Fig. 7.1 Geometric pattern to ore reserve estimation: **a** *Square* blocks for sampling or drilling design, **b** *Rectangular* blocks for irregular sampling design, **c** *Polygonal* blocks drawn by joining of mid-points between samples, **d** *Polygonal* blocks drawn by angular bisection, **e** *Triangular* blocks drawn by joining three adjacent sampling blocks, **f** Profiles through a hypothetical ore body

numerically change gradually and continuously along a straight line connecting two stations within the ore body. According to this rule, the value (mineralization parameter like thickness and grade) at any point between two adjacent stations is constant and equal to the value of the nearest station.

(2) Observations are made in conformity with the nature of the deposit and samples are taken with equal precision everywhere; such samples represent the ore within the zone of influence they control.

(3) That the physical continuity projected on the basis of exploration is a geological possibility in a given set up.

(4) An ore body of complex shape can be represented by drawing hypothetical figures, plans and sections.

(5) The advantages and disadvantages discussed in the case of the triangular method apply for this method also. An added disadvantage in this case is that the construction of proper polygons is a job which requires skill and experience.

7.2.1 Method of Reserve Computation

The United States Bureau of Mines has developed four methods for reserve computation, which are:

(i) **Average Factors and Area Method**

The method is considered as arithmetic average, average depth and area, statistical, analogous, geological block methods, etc. Here, we shall refer to them as geological block and analogous block methods. The assumptions here are that certain block or areas have geological and technical features similar to certain other blocks which have been explored or mined out and their results were known. The result of the known block is now taken as valid for an unknown block of similar dimensions. The blocks in such cases may be constructed on the basis of geology, structure, thickness, depth, grade, value and overburden.

(a) **Analogous Block Method**:

It is based on the principle that a quantitative similarity exists between a given block and another block which is better known (mined out). The variables such as thickness and grade may be the arithmetic or weighted average of a number of observations like drill core data and pit data. A single observation in the given block of these parameters is also acceptable.

The method is accurate in uniform deposits, the accuracy decreasing with the irregular deposit. This method is suitable for computing district-wise reserves on the basis of approximate parameters like tonnage/km^2 when an ore body is inadequately explored. It is a recommended method.

(b) **Geological Block Method**:

A geological block may be an entire mineral deposit of a small portion outlined in a map on the basis of exploratory data. The block can be defined by fixing boundaries on the basis of geological and structural features (fold, fault) or on the basis of variation in grade and thickness. Variables like grades and thickness are computed from production data, exploratory data, sampling data or data from another part of the ore body. The average grade and thickness are determined by arithmetic or weighted average or by statistical analysis. This method is effective in the early stages of exploration, the accuracy increasing with the increasing availability of exploratory data. Bedded or placer deposits are amenable to this method. The major advantage of both the methods is their simplicity, ease of direct application and relative accuracy in circumstances of insufficient data.

(ii) **The Mining Block Method**

This method is also known as longitudinal section method and mine exploration method. A mining block may be defined as an ore block bounded on four sides by openings/workings or bounded on three or less sides by openings/workings and other sides by arbitrary lines. Such blocking can be done on the basis of exploration (borehole sampling), geological features, and technical and economic considerations. Mining blocks may be rectangular in shape with their base lying in the plane of inclined or vertical longitudinal section.

(a) **Block Exposed on All Four Sides**

The area is computed by the length and breadth of an individual block and the thickness is determined as the arithmetic average of the assay width. When the ore thickness is less than the underground openings and uniform, the variables are found by calculating the simple arithmetic average. In an irregular ore body, the average grade of a working is computed by the weighted average method. The weight factor can be area, volume or tonnage. When the interval of sampling on all sides is equal, a simple arithmetic averaging will be sufficient for obtaining the average thickness and grades. If the interval of sampling or the lengths of the sides are unequal, thickness and grades are not directly computed. Here, it will be necessary to find the weighted average thickness and grade. The weight factor may be the lengths of the individual blocks. If the lengths of the sides are unequal and grades and thickness show wide variations, the average factors need to be found by working out the weighted averages. The weight in this case is the area of influence of the sample points. If the ore thickness is more than the width of underground openings and the blocks are developed by cross-cuts on two levels, the cross-cut data are taken for reserve computation where mineralization is continuous. In an irregular or discontinuous ore body, the blocks are subdivided and the areas of influence are computed separately for each subdivided block.

(b) **Block Exposed on Three Sides**

(i) The method is used effectively; arithmetic average of three sides, weighing the variable against the length of workings on each side of the block; (ii) computing the factors for the fourth side by and samples then averaging the variables on all sides; and, (iii) weighing the areas of influence of three existing workings. In case of an area blocked out by an adit and surface workings also, the above methods are used. Geological evidence like the degree of ore alteration, thickness, grade, zoning or the number of observations helps in defining the ore boundary, more so if the ore is exposed on the surface.

(c) **Block Exposed on Two Sides**

When a block of ore is defined only by two levels, the average factors for both the levels are computed separately. The block averages are worked out as the average of both the levels in uniformly mineralized ore bodies. If the ore is not uniform, the area weights are used for computing the weighted averages.

(d) **Block Exposed on One Level and Intersected by Borehole at Depth**

In this case, from the drill hole intersection levels perpendiculars are drawn to the working levels. Blocking of boundaries can also be done by geological criteria or by economic considerations.

$$\text{Average thickness } T_{av} = \frac{t_1L_1 + t_2L_2}{1_1 + 1_2}$$

$$\text{Average grade } C_{av} = \frac{C_1L_1 + C_2L_2}{L_1 + L_2},$$

where

t_1 average thickness of ore in each block in the level,
t_2 average thickness of borehole interactions of adjoining boreholes,
C_1 average grade of ore in each block in the level corresponding to t_1,
C_2 average grade of ore in borehole intersection of adjoining boreholes corresponding to t_2,
L_1 and L_2 block length on both the levels,
T_{av} average block thickness, and
C_{av} average block grade

When a large number of levels and boreholes are available, the average grades and thickness are computed by using the following formulae.

$$T_{av} = \frac{t_1 + t_2 + \cdots + t_n + t_2^1 + \cdots . + t_m}{n + m},$$

$$C_{av} = \frac{c_1 + c_2 + \cdots + c_n + c_1}{n + 1},$$

where

$t_1, t_2 \ldots t_n$ thickness observed in the drift,
$c_1, c_2 \ldots C_n$ corresponding assay values,
$t_1, c_2 \ldots Cn$ thicknesses of drill hole intersections,
$c_1, c_2 \ldots C_m$ corresponding assay values
n number of samples in the drift, and
m number of drill holes

For adopting this method, the ore body is blocked out by a network of openings and the block reserves are computed on the basis of the factors: (i) thickness, (ii) grade and (iii) mining costs. The degree of error shall depend on the genetic type of the ore. This method is highly effective except in highly irregular pocket deposits

(iii) **Cross-section method**

In this method, the ore body is divided into various segments by transverse cross-section lines spaced at equal intervals or in some cases at unequal intervals. The ore bodies are divided by such lines in different levels including underground workings. Three methods have been recognized.

(a) Standard method

Each internal block is confined by two section lines and an irregular lateral end. The end sections are controlled by one cross section line and one irregular end surface. Sections may be parallel or non-parallel, as well as vertical, horizontal or inclined. The reserves are computed as follows.

The areas of all sections are determined first. The average thickness and grade for each section are computed next. Thus, the volume of a block confined by two adjoining sections is

$$\text{Volume of ore } V = \frac{(S_1 + S_2) \times L}{2},$$

where S_1 and S_2 are areas of two adjacent blocks of nearly equal size and L is the perpendicular distance between them. The volume is multiplied by the specific gravity for computing the tonnage. Sample points can be as close to the section lines as feasible.

The end area of an ore body may occupy a number of irregular shapes, the areas of which cannot be determined easily. In such special cases, different formulae are necessary to compute the volume of the end area. There are also cases when S_1 and S_2 vary in size but are similar in shape. Some of the cases and formulae are briefly listed below.

More complex shapes and sizes will also be available in some cases which will have to be tackled on the basis of more sophisticated formulae.

(b) Non-parallel sections

Because of the change in the strike of the ore body, some of the transverse sections may show a tendency to diverge or converge. They are non-parallel and reserve computation becomes complex. If the sections are AA′, BB′ and the areas of cross sections involved are S_1 and S_2 and the angle between AA′ and BB′ is less than 100 and the volume is defined as

$$V = \frac{(S_1 + S_2)}{2} \frac{(h_1 + h_2)}{2},$$

where h_1 and h_2 are two perpendiculars drawn from the centre of the cross section to the adjacent section line. The centre of gravity of the section is determined by plotting out the area on a piece of cardboard and locating the point where the cardboard hangs in balance from the hanging thread.

Where the angle is more than 10°, the above formula is modified to

$$V = \frac{\alpha}{\sin\alpha} \frac{(S_1 + S_2)}{2} \frac{(h_1 + h_2)}{2}$$

(α to be expressed in radians)

(c) Linear method

Here, each block is defined by a section and a length equal to half the distance between the adjoining sections. Volume is determined by the formula $V = S_1 \times L$.

Well-defined, large ore body with uniform thickness and grade generally gives accurate results in cross-sectional methods. It the ore distribution is highly erratic and the ore body has an irregular shape, the method may not give satisfactory results. It is particularly useful in evaluating boreholes drilled on tabular ore body and obtaining intersections at different levels in a vertical plane.

(d) Method of isolines

Isolines (isopachs, isochors and isograds) are used for reserves computation by combining the plan with the principle of cross sections:

$$\frac{(S_1 + S_2)}{2} \times L = V$$

It is used for finding the volume between any two cross sections when the ore body is regular. More sophisticated formulae are required for complex situations where the value distribution is in two parts of the cross section. If the area corresponding to thickness "h" is in two parts ($S2'$ and $S2''$), the volume of the slice between h_1 and h_2 will be given by the formula

$$V_2 = h\left[\frac{S1 + (S2' + S2'')}{2}\right]$$

In a case where the ore thickness, say h_2 is missing in one section and h_1 is available then the slice between h_2 and h_3 is given by the formula:

$$V_3 = \frac{h}{2}[(S2' + S2'') + (S3' - S3'')]$$

In both the formula,

V volume,
h a constant thickness interval between the two isolines,
S_o area enclosed by ho contour line, and S_1, S_2, etc., are areas corresponding to the thickness h_1, h_2 contour lines

Isograd maps can be used for computing the average grades by the following formula:

$$C_{av} = \frac{C_0A_0 + \frac{C}{2}(A_0 + 2A_1 + 2A_2 + \cdots + A_n)}{A_0},$$

where

C_0 minimum grade of ore,
C constant grade interval between the isolines,
A_0 area of the ore body with grade C_0 and higher,
A_1 area of the ore body with grade C_0 plus C and higher and
A_2 area of the ore body with grade C_0 plus $2C$ and higher, etc.

Isoline maps have some advantage over other plans. They depict the ore body showing the distribution of poor and rich ore value. They are easy to read, measure and interpolate and in computation, the total calculations are not very large as there are fewer blocks unlike other cross-section methods. These plans also help in a better mine planning. However, it suffers from certain drawbacks. From the same set of data, inexperience evaluator may construct isolines with different configurations. Also, where the data are scattered, isolines may give a very erroneous impression of value distribution. In case of abundant data, the process of construction of the isoline maps becomes cumbersome.

(iv) **Analytical methods**

(a) Method of triangles

Here, all workings, pits or boreholes or any other exploratory openings are connected to one another by a series of straight lines. This divides the ore body into a series of triangles. For computational accuracy, equilateral triangles are preferable, but in practice the construction of such triangles is not feasible. Each triangle rests on the plane of the map but represents an imaginary prism with edges t_1, t_2, t_3 with a thickness equal to the vertical thickness of the ore body. The upper base of the prism is conceptually truncated (smoothened to an even shape) because of the irregularity of the ore body. The volume of a truncated triangular prism with uneven height is found out by the formula.

$$V = \frac{1}{3}(t_1 + t_2 + t_3)S,$$

where S is the area of the base.

The area S is computed by using the standard formula ½ base X height. It is preferable to use a common base line for a pair of adjacent triangles. The average grade of each prism is determined by the formula.

$C_{av} = \frac{C_1 + C_2 + C_3}{3}$ when the ore distributions are uniform. In case the ore distribution is erratic, weighted averages are found and the formula becomes $C_{av} = \frac{C_1 t_1 + C_2 t_2 + C_3 t_3}{t_1 + t_2 + t_3}$

From the volume, the reserves are computed by multiplying it by specific gravity. This method is used in deposits of sedimentary types, particularly the simple bedded types and large disseminated types explored by a good network of boreholes or pits. The method has limitations. The triangles do not permit delineation of the physical ore types. They may also camouflage value distributions. The method requires the subdivision of the ore body into a large number of triangles of which means more calculations. In lenticular ore bodies, this method shall possibly underestimate the total reserves.

(b) Method of Polygons

This is also known as the area of influence method. The procedure is to determine the area of influence of each exploratory point (pit, borehole) and construct polygonal blocks with the pit or borehole in the centre of each polygon. The area of influence is determined by joining two adjoining pits and finding out the midpoint between them. A perpendicular is dropped at this point of division. The process is repeated for all sets of boreholes or pits. The perpendiculars intersect and the polygonal forms automatically result. A correctly constructed polygon will have angles less than 180 between any two sides.

Difficulty comes in linear workings and horizontal boreholes, trenches. In such cases, the linear workings need to be divided into four areas of influence or four elementary prisms each characterized by appropriate workings. This is comparable to the mining block method. For computing the reserves, the area of the polygon is determined and is multiplied by the thickness of the ore within the polygon and the specific gravity of the ore.

This method is applicable when there are a large number of workings/exploratory openings arranged in a regular grid. Reserves of tabular bodies, large lenses and stocks are effectively computed by this method.

7.2.2 Criteria for Selecting a Reserve Computation Method and Classifications of Resources

The selection is based on the geology of the deposit, exploration method used, availability of data, reliability, purpose of calculation and the required precision. During preliminary estimates, any simple method can be used. But when the reserves need to be categorized for mine planning, a method compatible with the extraction technique needs to be used. In open cast operations of large uniform ore body, any simple method will give reliable results. But if the mineral is of low bulk and the values are erratic, and the ore has to be extracted by underground mining very precise methods need to be selected. The type and extent of exploratory openings also influence in method selection.

No reserve estimate is fully perfect. Within the inherent and well known limitations involved in the work, the estimate provides certain understandable precision. The accuracy of an estimate depends on the quantum and quality of the exploratory data. The more the exploratory data the better the precision of the estimate. Ore reserves need to be classified according to a decreasing order of precision. Accordingly, a classification of four types of reserves namely, developed reserves, proved reserves, probable reserves and possible reserves is most acceptable.

(i) Developed Reserves:

These are reserves blocked out and ready for immediate extraction, their quantity and grade are estimated on the basis of data from mine development carried out as preparatory to production.

(ii) Proved Reserves:

These include such reserves where the ore body limits and average characteristics are sufficiently assured and changes of failure of these estimations are so remote that a decision for mine development can be taken.

(iii) Probable Reserves:

This category of reserves includes either extensions around the proved ore panels or such zones where the ore body limits and average characteristics are known with reasonable certainty.

(iv) Possible Reserves:

Assumes the continuity of the ore body purely on geological grounds, past mining activity, comparison with similar deposits hereby, small-scale regional mapping with wide spaced drilling ore geophysical and geochemical evidences.

Yet another category of reserves known as “Potential Ore” can be added. Such ore is essentially submarginal and likely to become exploitable when technology and economic would support.

Reserves can also be classified in different contexts grade wise, depth wise, thickness wise, etc., in specific mining situation. However, the classification into "developed", "proved", "probable" and "possible" categories alone is helpful in making investment decisions. Grade-wise classification is a flexible concept and is used to define ores already classified into developed, proved grades. Here, size specifications are used along with grade, so that one refers to, say, developed reserves of +64% Fe and +10 mm size, in the case of some iron ores.

In addition to the classifications discussed so far, there could be others. The meaning of the term "reserves" differs from one person to another, i.e. geologist, miner, metallurgists and economist. While a miner works with a property for immediate returns, the economist likes to view on the future, not forgetting the current needs. Irrespective of the long- or short-term economic considerations, the computed reserves are usually geological reserves, computed by geologists.

The geological appraisals vary from organization to organization. In practice, we find no agreement between any two methods or any two organizations. Thus, neither standard classification nor norms of reserves calculation for any deposit are fully adopted. Therefore to overcome such an anomaly, decision takers are engaged defining terms with suitable definitions for expressing reserves, fixing certain boundaries beyond which these terms become invalid. From such attempts have emerged various sets of classifications like positive, probable and possible; measured, indicated and inferred. The Indian Bureau of Mines (IBM) some time ago had suggested the recognition of nine types of reserves in the appraisal of specific properties and four types of reserves in the appraisal of specific properties and four types in the appraisal of national resources. These are as follows:

Appraisal of specific mining properties as proposed by IBM:

(i) Minable proved reserves
(ii) Marginal proved reserves
(iii) Submarginal proved reserves
(iv) Minable probable reserves
(v) Marginal probable reserves
(vi) Submarginal probable reserves
(vii) Minable possible reserves
(viii) Marginal possible reserves
(ix) Submarginal possible reserves

Appraisal of Mineral Resources

(i) Minable measured resources
(ii) Marginal measured resources
(iii) Submarginal measured resources
(iv) Inferred resources

The classification of reserves as measured, indicated and inferred, as given by United States Bureau of Mines is widely used. It is proposed that the terms "measured" and "indicated" be replaced by a single term "demonstrated ore".

We agree that the classification of reserves is difficult similar to definition of "ore". Work on reserve estimation gives the amount and grade of lower ore for any particular time. It must also show the amount of ore available, even of a lesser grade that can be utilized profitably when technology is available. The total quantity of ore in the above categories, viz. reserves, marginal resources, submarginal resources and latent resources which together constitute "resources" need to be estimated. The parameters adopted by exploration organizations may differ according to the objectives, and the purpose of the assignment, but it needs to show comparison of the data and results.

"Identified resources—specific bodies of mineral bearing material whose location, quality and quantity are known from geologic evidence supported by engineering measurements with respect to the demonstrated category". "Undiscovered resources—unspecific bodies of mineral bearing material surmised to exist on the basis of broad geologic knowledge and theory".

Therefore, reserve is that portion of the identified resource from which a usable mineral can be economically and legally extracted at the time of determination. The term ore is used for reserves of some minerals.

The reserves identified are:

(1) Measured/Proved: Material for which estimates of the quality and quantity have been computed, within a margin of error of less than 20%, from sample analysis and measurements from closely spaced and geologically well-known sample sites.
(2) Indicated/Probable: Material for which estimate of the quality and quantity has been computed partly from sample analyses and measurements and partly from reasonable geologic projections.
(3) Demonstrated: A collective term for the sum of materials in both measured and indicated resources.
(4) Inferred/Possible: Material in unexplored extensions of demonstrated resources for which estimates of the quality and size are supported by geologic evidence and projections.
(5) Identified (subeconomic resources): Materials that are not reserves, but may become so as a result of changes in economic and legal conditions.
(6) Para marginal: The portion of subeconomic Resources that (a) borders on being capable of production economically or (b) is not commercially available solely because of legal or political circumstances.
(7) Submarginal: The portion of subeconomic resources which would require a substantially higher price (more than 1.5 times the price at the time of determinations) or a major cost reducing advance in technology.
(8) Hypothetical Resources: Undiscovered materials that may reasonably be expected to exist in a known mining district under known geologic conditions. Exploration that confirms their existence and reveals quantity and quality will permit their reclassification as a reserve of identified subeconomic resource.
(9) Speculative Resources: Undiscovered materials that may occur either in known types of deposits in a favourable geologic setting where no discoveries have been made, or in as yet unknown types of deposits that remain to be recognized.

Exploration that confirms their existence and reveals quantity and quality will permit their reclassification as reserves or identified subeconomic resources.

There is United Nations Framework Classification (UNFC) is also available for adoption.

7.2.3 Grade Computation

The grade of the ore defines the quality of the ore or in short the quantity of the valuable product within the ore expressed in terms of percentage availability. Four types of grades are usually recognized.

(i) Sample grade: The quality of the ore in situ as determined by underground, surface or drill hole sampling.
(ii) Mill-head grade: The grade of the ore as it comes out of the mine and sent to the mill.
(iii) Recoverable grade: This is the quantity of ore that can be recovered by the metallurgical process at mill after due losses. Some amount of ore is usually lost during the extraction. To account for this, the concept of recoverable grade is essential.
(iv) Liquidation grade: This is the grade of ore on which a purchaser or smelter fixes the price.

In addition, there are terms like cut-off grades, average grade, etc., which define certain specific conditions of the ore quality. The cut-off grade can be roughly defined as the lowest grade of ore which can be mined economically. Sometimes a given cut-off limit may coincide with some geological boundary. In such cases, the term geological cut-off can be used. Pay value is defined as the grade of ore that can be mined in any given techno-economic set-up. In computing reserves and average grades, the data should be assembled on the basis of cut-off grades. In borehole intersections, pit sections, channels, etc., the assay value will be distributed in such a way that some will be below the cut-off grade and some will be above. Those which are below the cut-off have to be eliminated before the average grade is computed. However, the rejections should be done with a great deal of caution. The following procedure is advocated. The average grade with all the values is calculated. Then the average grade is computed after eliminating the low values. If the difference is not appreciable, then, the lower values can be included. If they bring in a sharp dilution so that the average grade is near the cut-off grade, then the low values will have to be omitted. In case, the low value portions form a column of ore which is recognizable during mining, then that portion should be omitted from calculation in any case. It is easy to reject a leaner section only when manual mining and hand sorting operations are used.

The grade worked at any time is commensurate with the then current economics and technical knowledge. With improved technology, low-grade ores might be used later. This point should also be kept in view while computing the grades.

The average grade defines the quality of the ore of a well-defined block of ore. The computation of average grade is essential for classifying the reserves. Some of the methods for calculating the average grades are discussed below. Some concepts about the precision of grade estimates are also discussed.

The average grade should be computed on the basis of assay plans or isograd maps. In the simplest form, the average grade can be calculated by arithmetic averaging.

$$G_{av} = \frac{\sum a_1 + a_2 + a_3 + \cdots a_n}{n},$$

where a_1 a_2 … a_n are grades of individual samples in a block and n is the number of samples. However, the situation described above is extremely simplified and is applicable only when the ore is uniform in every respect which is seldom the case. Some non-uniformity is always present and to account for the influence of non-uniformity, the concept of weightage is introduced. Here, ore thickness intersected by a hole/pit, area of influence of a sample or the tonnage over the influence of the sample.

In cases where there is a question of minimum stope width, the weight will always be the stope width with allowances for dilution. In such cases, the principle of weightage has to be employed even the individual channel samples. The minimum stope width is $x + y + z$, where $x = 1.0\%$ Cu, and $y = 2 = 2.0\%$ and $z = 0.5\%$ Cu. In finding out the average grade, the stope calculation will be:

$$G_{av} = \frac{X \times 1.0 + y + 2.0 + z \times 0.5}{x + y + z}$$

The x, y and z are the weighting factors which in this case, are widths. The weighting factor can be area or tonnage, also, as indicated earlier.

Precision of Grade Computation: The average grade of ore is generally expressed as say $X\%$ Al_2O_3, Fe or any other valuable mineral content, or as 2% Cu, 4 gm of gold, etc. The value of "*x*" is determined by either arithmetical average or weighted average. The expression that "*x*" is the average grade of ore is known as a point estimate. Such an estimate generally implies an accuracy unattainable in most mineral deposits. Hence, the grade of ore should be expressed within a range of values, say 64–66% Fe. With this, a statistical confidence interval is also defined and the estimate becomes more acceptable in the light of fluctuations easily noticed in the grade values. Confidence interval defines the lower limit and upper limits within which most values will fall in a sample from a panel, level, mine or a set of boreholes. The confidence interval generally in use is 95% or to 0.5 which means that in 95 cases out of 100, the values will be within a set of values defined by the confidence interval. For an estimate of lesser precision a 90% confidence limit is also used. The calculation of the confidence limit for one set of data is shown in the case of bauxite deposit in the chapter on statistical methods. An estimate with confidence intervals achieves objectives which guide to arrive at best decision.

Keeping in mind the following: (i) The risk inherent is known quantitatively, (ii) The fluctuation in grade can be understood, (iii) By increasing the sampling the precision of estimate can be improved. Confidence interval helps in knowing the improvement in the accuracy of estimate quantitatively, (iv) Cut-off can be determined more realistically.

Further Readings

Arogyaswami RNP (1988) Course in mining geology, 3rd edn. Oxford IBH, p 695

Buibbert JM, Park CF (1986) The geology of ore deposits. WH Freeman, New York, p 985

David JB (1977) An introduction to geographic information systems, The GIS Primer. www.innovativegis.comfbash/primerrrhe_GIS PrimerBuckley.pdf, 1997, p 215

Deutsch CV, Ross ME (2013) Mineral resource estimation. Springer, Berlin

Gocht W, Zantop H, Eggert RG (1988) International mineral economic. Springer, Berlin, p 271

Guilbert JM, Park CF (1985) The geology of ore deposits. WH Freeman, New York, p 985

Harris DP (1984) Mineral resources appraisal. Oxford University Press, Oxford, p 445

Kearey P, Vine FJ (1992) Global tectonics. Blackwell Scientific, p 302

McKinstry ME (1960) Mining geology. Asia Publishing House, Prentice Hall, p 680

O' Hara TA (1980) Quick guide to the evaluation of ore bodies CIM. Bull 73(2):87–99

Onle EL, Bates RL (1981) Geology, geologists and mineral exploration. In: Skinner BJ (ed) Economic geology 75th anniversary, pp 775–805

Rendu JM (1978) An introduction to geostatistical method of mine valuation. S Afr lnst Min Metall Monograph Ser Geostat 2:84

Rose A, Hawkes HE, Webb JS (1979) Geochemistry in mineral exploration. Academic Press, p 657

Sinha RK, Sharma NL (1993) Mineral economics. Oxford and IBH Publishing Co. Pvt. Ltd., p 394

Wellmer FW (1989) Economic evaluations in exploration. Spring, p 163

Chapter 8
How to Arrive at a Decision to Open a Mine?

8.1 Exploration of Mineral Deposits: Field Guides and Preliminary Observations

Each mineral deposit has characteristics of its own; the deposits of same mineral occurring close cluster show diverse patterns. Yet broad similarities of certain characteristics are shown which patterns guides to inferences. On this basis it is important to recognize field guides which help in locating a given mineral deposit.

It is understood that no two deposits, even though of the same mineral or ore, bear absolute comparison with each other; an exploration plan successful in one deposit may not meet success in another. In planning an exploration strategy, an understanding of the geological knowledge and other data of each deposit is necessary. In order to collect data systematically, the following checklist of observations to be made and recorded is recommended. A checklist of preliminary observations is given in Table 8.1.

These factors help in selecting exploration method and the quantum of exploration. The latter is dependent on the required degree of precision of the exploration data.

After a promising mineral deposit has been located, its broad characteristics studied, the next step is to choose a scheme for detailed exploration for specific needs and aspect of cost, time. Such considerations help the exploration geologist in choice of an exploration method. They may be made used either individually or in combinations.

1. **Obtain Reliable Type of Samples in Sufficient Quantity**

The aim of any exploration programme is to collect samples of required quantum and type in order to prove the extent of the deposit. Hence, it is important to choose a method which yields such samples laterally and in depth. Samples need to be available continuously at regular intervals. In the surfacial deposits, pitting yields most reliable samples. But, as mineralization extends downwards, most of the

G.S. Roonwal, *Mineral Exploration: Practical Application*,
Springer Geology, DOI 10.1007/978-981-10-5604-8_8

Table 8.1 Field guides for preliminary observation in an exploration camping

Status of the area	(a) Developed area with good communications, or (b) Undeveloped area	Wall rock alteration	(a) Pronounced, (b) Prominent, (c) Not prominent of (d) Totally absent
Topography of the terrain	(a) Plain/wooded/barren (b) Hilly/wooded/barren (c) Mountainous/wooded/barren	Mineralogy	(a) Monomineralic, (b) Polymineralic, and (c) Nature of gangue minerals
Regional geology and structure of the terrain	(a) Igneous, (b) Sedimentary, (c) Metamorphic, (d) Residual, (e) Alluvial terrain, (f) Structure and tectonics-sample or complicated	Structure	(a) Simple, (b) Fairly complex, (c) Complex, or (d) Very complex
Size of the ore body	(a) Length, (b) Width, and (c) Depth	Disposition	(a) Extent of overburden (i) No overburden, (ii) Partially covered by overburden, or (iii) Fully covered by overburden (b) Depth extension (i) Confined to the surface, (ii) Partially confined to the surface, or (iii) Fully extending underground
Outline of the ore body	(a) Simple, irregular, or (b) Complex	Ground/surface water	(a) Present, but not likely to cause any trouble, (b) Present and likely to interface with drilling, pitting, trenching, aditing, etc. or (c) Not present at all
Types of ore bodies	(a) Capping, (b) Reef, (c) Veins (various types of veins), (d) Disseminations, (e) Lenses and pockets and other irregular bodies, or (f) Stratified sedimentary beds	Field evidence about ore genesis	(a) Clear, or (b) Not clear
Nature of the enclosing rock	(a) Lithology, (b) Hard, medium, soft, friable, etc.	Climate of the area with records of rainfall and temperature variations	(a) Desert, (b) Arid, (c) Semiarid, (d) Humid, etc.

observations have to come from diamond-core drilling which can later be substantiated by samples obtained from exploratory mining.

2. **Efficiency of the Method in Operation**

This depends on a host of conditions like the organization of the team, nature of the country rock, stratigraphic sequence, nature of mineralization, structural features, etc. Pitting in hard rock like quartzite is time-consuming and costly and it can be resorted only under compelling circumstances. In such cases, core drilling will yield equally reliable data.

3. **Financial Affordability**

The total cost of operations should not become unbearably high. The geologist should remember that the money sunk in exploration cannot be recovered in any form except in the sale value of the ore to be proved. The amount expended on this account should be such as to be easily absorbed by the sale price later.

4. **Maintenance of Required Speed of Operations**

The exploration programmes are mostly time bound. Sometimes, either due to breakdown of machinery or any other reason, the programme may get stalled. Hence, the method chosen should be quite flexible with either spare machinery or manpower, or by opting for substitution of the programme with another suitable method.

5. **Conformity with the Shape, Size and Pattern of Mineralization**

This factor almost totally controls any exploration operation. For deposits of large aerial extent and surfacial nature, pitting and trenching are adopted. Asbestos, bauxite, magnesite and diamond are explored by these methods However, spacing of the pits or trenches is of vital consideration as it influences the total cost of operation. Care should, however, be exercised in exploring some minerals like mica either by pitting or by drilling. For deep-seated deposits such as chromite, copper, lead and zinc ores, drilling will have to be done. Drilling in such deposits is often controlled by the structure and stratigraphic sequence.

6. **Topography of the Area**

Some areas remain inaccessible due to their altitude, thick vegetation, poor ground condition like marshy terrain and lack of communication. Though man can reach these areas, it would be very difficult to take material and machinery. This particular situation has been faced in the case of bauxite deposits. Even many iron ore deposits had been inaccessible in India until recently. The exploration geologist will have to choose such a method which entails easy transportation of man and material, availability of firm ground for setting up camps and drilling machines, availability of other facilities, etc.

7. **Marketability of the Ore at the Exploratory Stage**

In some exploration programmes, it may be necessary to sell the ore raised during the operation. In such cases, the choice may be in favour of a method which may yield large quantities of ore. This is generally the case in the exploration of copper ore which is sent to smelters even far removed from the deposit. Similar is the situation with mica, asbestos, etc.

8.2 Important Mineral Deposits and Exploration

India possesses a large number of metallic and non-metallic mineral deposits which are found catered throughout the length and breadth of the country. Of these, the most important deposits are chromite, asbestos, nickel, magnesite, diamond, barites, fluorite, copper, lead-zice, gold, tin, tungsten, mica, limestone and dolomite, gypsum, phosphorite, iron ore, manganese ore, kyanite, sillimanite, graphite, talc, pyrophyllite, bauxite and clay.

For convenience of discussion, these mineral deposits have been grouped on the basis of their host rock associations as given below

(1) Mineral deposits associated with igneous ultramafic and mafic rocks.
(2) Mineral deposits associated with igneous intermediate rocks.
(3) Mineral deposits associated with igneous acidic rocks.
(4) Mineral deposits associated with sedimentary evaporate rocks.
(5) Mineral deposits associated with metamorphic rocks.
(6) Mineral deposits associated with residual rocks.
(7) Mineral deposits associated with placers.

8.2.1 *Mineral Deposits Associated with Ultramafic and Mafic Igneous Rocks*

The major ultrabasic and basic rocks are peridotite, dunite, saxonite, pyroxenite, enstatite, norite, dolerite, basalt, gabbro, etc. These rocks may be extrusive or intrusive in nature. The important mineral deposits associated with these rocks are chromite asbestos, magnesite, platinum, titaniferous magnetite, corundum nickel-copper sulphides, silver, cobalt, etc. Of these, the most important deposits are chromite, asbestos, nickel, magnesite, diamond, barites and fluorite. The geology, prospecting and exploration of these deposits are dealt with below.

(I) **Chromite**

Chromite deposits occur mostly in the Precambrian ultrabasic intrusive. Five types of commercially important deposits are recognized. They are

(1) Evenly scattered
(2) Schlieren banded
(3) Stratiform (with bedded appearance)
(4) Sack form (pocket), and
(5) Fissure form (vein-like)

Chromite is a product of magmatic segregation and is invariably associated with intrusive ultrabasic igneous rocks. All the Indian deposits are considered as primary differentiates of ultrabasic magmas which intruded successively, producing several generations of chromite deposits. The earliest phase of the intrusion gave rise to coarse-grained chromite. This was followed by medium- and fine-grained deposits. Many ultrabasic intrusive, particularly in Orissa, show extensive lateritization which has remobilized the disseminated chromite, producing extensive lateritic deposits. Some chromite ore bodies show signs of genetic association with shear zones. This is particularly evident in Orissa and Karnataka deposits where they occur very close to zones of mylonitisation. Some ore bodies, particularly those of Keonjhar area, show signs of having been folded along with their host rocks, producing complexly folded ore bodies. The ore bears genetic association with region synclinal troughs.

Prospecting and Exploration
The important field guides for locating chromite ore bodies are shown in Table 8.2. Since chromite deposits occur exclusively in ultramafic rocks, the primary aim of prospecting is to locate such targets. Chromite-bearing ultramafic rocks are easily recognized in the field. The target area may be mapped on a scale of 1:25,000 to separate promising ultrabasic rocks which show obvious evidence of chromite mineralization. Outcrops to chromite may be usually present and may be of mappable dimensions. In case mineralization is suspected but outcrops are not visible, geochemical and geophysical prospecting may be carried out to locate specific deposits. Geophysical prospecting by gravity and electrical methods has helped in locating chromite ore bodies in Maharashtra and Orissa. When the densities of the ore body and the host rock differ sharply, the gravity method is successful. The ultrabasic bodies generally show contrastable magnetic characteristics which help in mapping them from their host country rocks. Electrical methods are useful in locating chromite ore bodies in serpentine zones because the electrical resistivity of chromite is 2–3 times that of serpentine.

(II) **Asbestos**

Asbestos is a name applied to a group of minerals of varying compositions. Asbestos has been classified mineralogically as follows:

(1) Amphibole group

(a) Orthorhombic

(i) Anthophyllite

Table 8.2 Exploration Criteria for Chromite deposits

S.L	Method	Stratiform deposits		Podiform bodies		Remarks
		Regional exploration stage	Intensive exploration stage	Regional exploration stag	Intensive exploration stage	
1	Mapping	1:2000–1:5000	1:1000–1:2000	1:2000–1:5000	1:1000	To isolabeultrabasics and mappable chromite bodies, sheer zones, zones of serpentinization, etc.
2	Drilling	200–500 m intervals	100 m intervals	100–200 m section intervals, 1–2 boreholes in each section	50–100 m section intervals at ¾ levels down to a workable depth	Also study core recovery in the ore zone, lithology, occurrence of unsuspected ore bodies, etc.
3	Trenching	As necessary to expose the concealed extensions	Not recommended	As necessary to explore the concealed sections along strike	Not recommended	–
4	Pitting	1–3 numbers across each representative ore type	3–5 numbers for every mass/body	2–4 numbers across for every representative ore type	3–5 numbers for every mass/body	More helpful in vein and lens-like bodies. Recovery with depth should also be studied
5	Sampling	Core and sludge, bulk and chips from pits, etc.	Core and sludge bulk samples for grade analysis and beneficiation	Core and sludge, separate analysis, for gangue-freen ore, bulk samples from pits according to ore types	Core and sludge samples from pits for grade analysis for beneficiation	–

(ii) Amosite-rich in iron than anthophyllite, and
(iii) Tremolite

(b) Monoclinic

(i) Actinolite, and
(ii) Crocidolite

(2) Chrysotile group

(a) Chrysotile, and
(b) Picrolite

Chrysotile asbestos is essentially a fibrous type of serpentine. Serpentinization is a type of autometamorphism which occurs in ultrabasic bodies like dunite. This autometamorphism is accompanied by emanations of hot residual solutions from the rock body itself. In the first stage, some 40–60% of the rock is serpentinized. In the second stage, the serpentinization takes place along fractures and openings. The fibre of asbestos is considered to be developed within the fracture zone.

Amphibole asbestos is considered to have originated due to deep-seated metamorphism. Molecular reorganization of the host rock without transfer of its constituents is thought to be the process responsible for the formations.

Geological structures have played a major role in the control and localization of some important asbestos deposits. Field evidences supporting this hypothesis are shown below

(1) Mineralization is richer within the synclinal folds, the secondary synclines showing the richest deposits.
(2) The noses of anticlines are devoid of asbestos mineralization.
(3) Where folding is not intense, mineralization is poor.
(4) The major ore shoots show pitches parallel to the enclosing structure
(5) When the synclines are shallow, yellow and light, asbestos are seen in greyish serpentine. Black serpentine is absent.
(6) In shallow synclines, the veins occur over a wider range of serpentinized zone, and do not remain localized as in the case of deep troughs.

Regional scale folds may also influence asbestos mineralization. Asbestos occurs in the form of very narrow seams and mass fibres. Individual veins may have lengths varying between 0.3 and 30 m and width ranging between 0.005 and 0.3 m. Individual fibre length may vary between a few mm and some 50 cm.

Prospecting and Exploration

Asbestos deposits of tremolite and chrysotile varieties are found only in ultrabasic rocks, their weathered derivatives, chilled contacts and magnesian limestones near the ultrabasic intrusive. Therefore, the area of initial search is exclusively confined to the terrain where ultrabasic rocks occur. As in the case of chromite, the ultrabasic rocks should be mapped out on suitable scales initially. The asbestos-bearing shear zones, or chilled contacts, can be mapped only if they are of large mappable dimensions.

Table 8.3 Exploration guides— asbestos

	Method	Details
1.	Mapping	1:200–1:600 in ideal host rocks
2.	Pitting	Random trial pits followed by systematic pits in 3–15 m grid and size varying from 1.5 m × 2.5 m × 2.5 m with variable depths, to ascertain grade and structure of the deposits
3.	Trenches	Shallow to deep, depending on the need but usually at 15 m intervals
4.	Drilling	15–30 m interval or at larger intervals, for regional exploration
5.	Exploratory mining	Drives, cross-cuts, winzes and raises to explore the contacts, and bulk sampling
6.	Sampling	Visual sampling to get ideas on fibre content

Airborne as well as ground magnetic surveys help in locating ultrabasic bodies which may be potentially asbestos-bearing. Although asbestos is nonmagnetic, it is associated with magnetite and this association makes magnetic methods effective. For aeromagnetic surveys, flights are arranged at 4 km intervals at a flight elevation of 150 m. In case the terrain is very rugged, the flight height is reduced to 90 m. Ground magnetic traverses are spaced at 60 or 90 m and readings are taken at every 15 or 30 m. Such surveys also can discriminate asbestos-bearing serpentinous horizons although actual ore zones cannot be distinguished this way.

The exploratory sequence followed in asbestos and other relevant details are shown in Table 8.3.

Visual sampling is a technique used in cross-fibre asbestos deposits, and gives reliable results about the fibre content within the ore zone without the aid of mining and milling tests. In this, all visible fibre exposed in slopes, drives and other openings, drill cores, etc. are counted and measured in terms of 1.59 mm fibre lengths as being useful and those below as unusable. The widths of all individual veins encountered are totaled up and the percentage of fibre to the width of ore is computed for each section. In a drive, the average value is calculated by the weighted average method. The values so obtained can be converted into percentage fibre per tonnes also. The length ranges of the fibres and their individual percentage availability for every tonne are also calculated. For conducting visual sampling, the fibres from individual veins, seams, etc. should be systematically and carefully scooped out by a pocketknife and measured by a scale. In case of partings within seams, fibres up to each parting should be scooped out separately. The widths of the individual seams are automatically fixed by their fibre lengths and this makes it easy for the weighted average computation for a certain drive length, stope or sample width.

(III) Nickel

Nickel deposits are known in two commercial forms.

They are: (i) Residual concentration of nickel silicate formed by the weathering of ultrabasic igneous rocks, and (ii) Nickel–copper sulphides formed by magmatic concentration and hydrothermal action.

Prospecting and Exploration

Prospecting for nickel in lateritic terrain is usually done by geochemical sampling. Nickel bearing laterites do not have any typical diagnostic characteristics which are different from any other laterite. Therefore, it is not possible to detect the presence of nickel in the laterite by mere physical examination. The presence of nickel in Sukinda/Odisha went unrecognized for many years although nickelliferous laterites were being stripped off for mining chromite. Grab samples taken rather casually from a yellow ochreous zone below the chromite deposit showed the presence of nickel for the first time. This discovery was followed by a systematic channel sampling effort in all exposed ochreous zones. All the samples indicated nickel in commercially viable quantities and the presence of a deposit was confirmed.

Geochemical subsoil and sub-outcrop samples are ideally suited for nickel prospecting and have been successfully employed in a few instances. In a nickel prospect in Orissa, the geochemical samples just below the topsoil showed the best nickel concentration. A similar experience is reported from Jamaica from a similar geological terrain. This is rather suggestive that the top layers of soil, laterities, etc. may not show the presence of nickel in recognizable and geochemically anomalous quantities. Geophysical prospecting may help in locating potential ore-bearing ultrabasic rocks.

The exploration for lateritic deposits is summarized in the section dealing with bauxite.

(IV) **Magnesite**

Three types of economic deposits of magnesite are known. They are

(1) Replacement in dolomite and limestone,
(2) As emanation, and
(3) Sedimentary beds

Prospecting and Exploration

The vein type of deposits is exclusively associated with ultrabasic rocks, particularly dunite. The area of preliminary search is narrowed down to areas of ultrabasic intrusive. Field guides for locating magnesite deposits are given in Table 8.4

The ultrabasic body suspected to contain magnesite may be mapped on scale 1:25,000 or 1:10,000. Individual mineralized patches are mapped if exposed and large enough. The presence of magnesite is fairly easy to recognize in ultrabasic

Table 8.4 Guides to the exploration of diamond

	Methods	Details
1.	Mapping	1:10,000–1:20,000
2.	Pitting	Widely adopted. No defined intervals
3.	Drilling	Useful in exploring vertical depth of pipes
4.	Expl. mining	Short levels with connecting raises and winzes
5.	Sampling	Pits and bulk samples from mines

rocks. The vein network usually shows bright grey to white colour in a greenish, dull grey background. Test for effervescence by acid on the white veins confirms the presence of magnesite. A few chip samples collected from important exposures may be analysed chemically and studied in thin sections for confirmation. Trenches may be put to study the extension of veins in depth. Scout boreholes to test the extent of mineralization and the continuity of the ultrabasic body in depth would be quite useful.

Prospecting for replacement deposits is rather difficult because dolomitic limestone and limestone which form the host rock and magnesite shows similar physical appearance. However, such deposits invariably occur in dolomitic terrain and the initial target can be confined to areas of dolomitic limestone. The initial mapping of the target area should be on a fairly large scale, 1:5,000–1:10,000. Systematic channel samples are collected at 50–100 m intervals. Their sections are prepared for samples at very close intervals, for every separate channel. From the microslides, the presence and percentage of dolomite is determined. The samples are subjected to chemical analysis in sections, corresponding to the channel sections from which microslides were made. A relationship between the magnesite availability shown by chemical analysis sand the corresponding microslide is established. With the help of the microscopy and chemical analysis data, the bands of magnesite and dolomite are marked out and mapped. Gradually, the prospecting geologist should acquire the necessary skill to separate dolomite and magnesite bands without difficulty.

(V) **Diamond**

Three types of diamond deposits are known

(1) Primary deposits in Kimberlite and other ultrabasic pipelike intrusive,
(2) Conglomerates, and
(3) Alluvium of stream and riverine origin.

Prospecting and Exploration

As mentioned earlier, all the three types of deposits occur in close proximity and the discovery of one often leads to the other two. The target area should generally be one which has yielded diamonds at some time and should be known to contain kimberlite intrusive.

The area may be mapped on large scale, 1:20,000–1:10,000 aerial photographs. The preferred targets are circular or elliptical water-filled depressions which may cover hidden kimberlite bodies and minor drainage channels and their alluvial accumulations for alluvial diamonds. The pipe-like kimberlite bodies generally weather easily and circular water-filled depressions are formed which are easily recognized in aerial photographs. Having chosen a large number of such targets, each water-filled depression is subjected to deep test-pitting to check the presence of weathered remnants of pipe or calcareous tuffa. This method was followed successfully by the Geological Survey of India in its Panna diamond prospecting.

Aeromagnetic, ground magnetic and electrical resistivity methods can be used for the prospecting of kimberlite pipes. The Hinota and Angora pipes of Panna area were discovered by these methods.

For placer diamonds, the preferred targets are minor confluences, alluvial fans and cones, river terraces, abandoned river courses, etc.

During ground verification, the targets selected on aerial photographs are checked by panning surveys. It, is, however, seldom can that diamonds be directly located by this method, because of their extreme rarity even within the pipe rocks. Diamonds are associated genetically with a number of satellite minerals. They are pyrope, picroilmentie, chrome-diopside and olivine. Since these satellite minerals occur in much larger quantities even within the pipe, their dispersion within the stream and alluvial accumulations is also proportionately more. Thus, the occurrence and concentration are carefully examined by panning survey. The panning points can be chosen from favourable targets located on the basic of photogeological studies. The points should be chosen systematically at regular grid intervals along current and ancient drainage channels till their source of origin or a source of origin of diamond is reached.

If diamond is located or any of the satellite minerals are seen in unusual concentrations during pan washing, the next step is to collect bulk samples of the gravel by deep pitting. This sample is subjected to washing and jigging to ascertain the presence of diamond. Here the fractions which are above 4 mm in size are straightaway rejected, and −4 mm +1 mm material is concentrated. The details of these investigations are systematically recorded in the geological map for choosing promising sites for detailed exploration. It may, however, be noted that, in the Indian context, the satellite minerals may be olivine, magnetite, hematite, limonite, perovskite and rutile.

The presence or absence of diamond in any terrain can be established only by bulk sampling and beneficiation of the samples, and forms the most important prospecting operation. In order to confirm that the beneficiation is complete and proper in every respect it is customary to add two–eight marked diamonds are not recovered the process of concentration is till the marked diamonds are recovered.

In India, diamond has been mined since a very long time and it is not unusual to find ancient workings. These workings can be found by ground reconnaissance and photogeological studies and form good field guides.

Alluvial diamonds may be found in two types of alluvium

(1) Placer deposits associated with the present drainage, and
(2) Placer deposits which show no relationship with the present drainage

In the first case, the deposit may be arranged linearly along drainage channels and therefore pitting, drilling, etc. may be done along line. In the second case, deposits tend to be irregular and a square grid system may be more effective in drilling and pitting.

Exploration guides for diamond is summarized in Table 8.4.

(VI) Barites

Three types of barite deposits are known in India. They are

(1) Vein deposits,
(2) Bedded deposits, and
(3) Replacement deposits.

Prospecting and Exploration
Barites mineralization tends to be associated with areas of uplift and magmatic activity of the granodiorite and admalite type. In the regional search for target, such areas should be given priority attention. Veins and beds of barites, when exposed, usually leave a solution creep because the outcropped material is dissolved away in water. This is a very important field evidence to look for. Float ores although not common, in some cases, may be found.

In order to define ground targets, geophysical methods are useful. In an investigation for vein barites in a granitic terrain, magnetic and resistivity measurements were used which indicated clear anomalies in areas of vein quartz with barites mineralization. This was followed by gravity surveys in which readings were taken at 5 m grid intervals which gave further definition to the ore zones.

Bedded deposits usually show genetic association with tuffaceous beds which may be useful as a field guide during prospecting.

(VII) Fluorite

Two types of fluorite deposits are recognized in India. They are (i) hydrothermal deposits associated with acid igneous rocks, and (ii) deposits formed in carbonate rocks, following the emplacement of alkaline igneous rocks, and associated explosive volcanic action.

The fluorite deposits of Madhya Pradesh and Rajasthan may be of hydrothermal origin whereas the deposits of Gujarat are in carbonatites and associated with explosive volcanism. The mineralizing solutions in this case are thought to have originated from the Deccan Lavas.

Major fluorite deposits all over the world occur along major continental rift zones and lineaments. The mineralization solutions are thought to have come up from the lower crust through the rift zone to form the mineral deposits. Some of the Indian fluorite deposits also occur along the Narmada rift zone of Tertiary age and may have a similar origin. The structural association in such cases is quite clear, the mineralization being broadly influenced by fault planes, as seen in the case of fluorite deposits of Gujarat and Madhya Pradesh.

Fluorite deposits occur generally in the form of veins and stringers within shear zones, and show considerable depth persistence, narrow outcrops and steep-to-vertical dips.

The stringers form a network coalescing and separating frequently. The individual ore lenses are of very small strike lengths. The Chandidongri deposit which is typically lensoid has a strike length of only 192 m, whereas the shear zone is traceable for a distance of 20 km. The deposits at Ambadongar also occur in the form of lenses of extremely small dimensions.

Prospecting and Exploration

In order to choose broad targets, photogeological studies of regional rift structures may help, as most fluorite deposits show genetic relationship with regional rifts. Areas of carbonatite intrusions, volcanism, radioactive mineralization, etc. are excellent targets. The presence of fluorine in ground water is a confirmatory evidence indicating the presence of fluorite mineralization nearby. Float ores are seldom seen. In a favourable target area, regional shear zones with quartz and calcite vein intrusive form favourable host rocks. For confirming the presence of fluorite, chip samples may be collected and analysed from outcrops. Gossan and cappings are usually present particularly with CaF_2 and are good indicators for the presence of fluorite mineralization. Geophysical prospecting has not been successful in locating ore bodies so far in India.

8.2.2 Mineral Deposits Associated with Igneous Intermediate Rocks

The intermediate rocks are diorites, nonozonites, granodiorites, syenites and nepheline syenites. The mineral deposits are essentially sulphides of copper, lead, zinc, gold, silver and iron. A large number of minor minerals are also recognized within this group of rocks. The important ones are molybdenum, nickel, cobalt, cadmium, selenium, tellurium, mercury, arsenic, etc. These may occur in complex sulphide form with several minerals or in single form with only one dominant mineral. In India, copper, lead, zinc, gold and pyrites are the most important ones.

(VIII) **Copper**

Copper deposits of many types are known, some of which are listed below

(1) Magmatic
(2) Contact metasomatic
(3) Hydrothermal
 (a) Cavity filling
 (b) Replacement
(4) Sedimentary
(5) Surficial oxidation
(6) Supergene enrichment

Prospecting and Exploration

Copper occurs in a variety of rocks, but most of the important commercial deposits in India are confined to the Archaean and Precambrian formations. All important copper mineralizations in India are associated with prominent narrow, linear greenstone belts and most of the deposits are either within the belts or in granites and gneisses in close proximity. These greenstone terrains therefore form the most important initial target area for search.

Copper has been mined in India since ancient times and old workings, which are very common, form the most important field guide. Old tailings, slagheaps, water-filled depressions with luxuriant growth of vegetation around should attract attention as a possible copper target. Such old workings can be easily located by photogeological studies and ground verification.

Geochemical sampling is very effective in defining areas of copper mineralization. The methods used are soil, bed rock and stream sediment sampling. For this, traverse and samples collected at every few metres on a regular grid pattern. Traverse lines at 100 m interval and sampling points at every 25 m along the line gave very promising results in prospecting for a copper deposit in MelanjKhand in MP.

Geophysical prospecting may be undertaken in areas where sulphide bodies are to be located. The methods may be airborne magnetometric surveys (for large regional targets), electromagnetic surveys and electrical methods (for specific small areas). In the Malanjkhand sulphide deposit, induced polarization (chargeability) gave very good results, particularly in pinpointing the area of mineralization. The prospect was earlier surveyed by self-potential and electromagnetic methods which also demarcated the area of sulphide mineralization. However, the anomaly was better defined by the use of induced polarization method.

(IX) **Lead-Zinc**

Lead and zinc deposits tend to occur together. The following types of lead-zinc deposits are recognized (Table 8.5).

(1) Contact metasomatic
(2) Hydrothermal
    (a) Cavity filling
    (b) Replacement
    (c) Disseminated

The known Indian deposits have been generally considered as hydrothermal, the mineralization being controlled by limestone and other calcareous host rocks. The geology and distribution of some important lead-zinc deposits are listed in Table 8.6.

Some lead-zinc deposits show evidence of having been controlled by lithology and structure in their genesis. The Zawar deposits occur within impure limestone (lithological control) and are also within a shear zone (structural control). The Sargipalle deposits also occur in an impure calcareous rock.

All lead-zinc deposits have been grouped as hydrothermal, although of late this assumption has been questioned. Some authorities point out that the deposits at Ambamata, Zawar and Sargipalle occur within geosynclinals sequence, as indicated

Table 8.5 General guidelines for exploration for sulphides (Copper, Lead, Zinc, Pyrites, Nickel (in sulphide) and multimetal combination)

Method	This, law dipping strate-bound body with a little structural formation		Large lonticular replacement body		Low to moderately dipping large bodies with simple, structure		Steap dipping small ore bodies structurally complicated		Remarks
	Regional exploration stage	Intensive exploration stage	Regional exploration stage	Intensive exploration stage	Regional exploration stage	Intensive exploration stage	Regional exploration stage	Intensive exploration stage	
1. Mapping	1:2000–1:5000	1:1000–1:2000	1:2000–1:5000	1:1000	1:2000–1:5000	1:1000–1:2000	1:2000–1:5000	1:1000	To establish foot-wall hanging wall contacts, extension along strike, individual leases and vein assay boundaries etc., lithological units, old workings, abandoned mines dumps, etc.
Underground mapping	1:100–1:200	1:100–1:200	1:100–1:200	1:300–1:200	1:100–1:200	1:100–1:200		1:300–1:200	
2. Drilling	200–500 a section intervals with 50–100 m grid drilling is test stripe	50–120 a grid drilling with underground drilling if necessary	200–240 a intervals in 2 levels within ore zone with 50–100 m grid intervals is test stripe	50–120 m section interval with intersection at ¾ levels at 50–60 a vertical interval with underground	240 m section interval, 50 m down depth within ore body to be explored	100–120 a section interval. Intersection at 3–4 levels with 50–60 m intervals, underground	50–100 m section interval, intersection at 25 m vertical depth in ore body	50 m section interval, intersection at 4–5 levels at vertical intervals of 30 m	For establishing depth continuity, strike extension, strike extension, width of mineralization and depth of oxidation

(continued)

Table 8.5 (continued)

Method	This, law dipping strate-bound body with a little structural formation		Large lonticular replacement body		Low to moderately dipping large bodies with simple, structure		Steap dipping small ore bodies structurally complicated		Remarks
	Regional exploration stage	Intensive exploration stage	Regional exploration stage	Intensive exploration stage	Regional exploration stage	Intensive exploration stage	Regional exploration stage	Intensive exploration stage	
				drilling if necessary		drilling if necessary			
3. Exploratory mining	Not recommended	1 k. strike length of the deposits by driving cross-cutting etc.	Not recommended	The entire strike length is two levels by drives, winses, cross-cuts	Not recommended	Entire strike length is two levels by drive with cross-cuts at 30 a intervals and winses and rainses	Not recommended	Entire strike length in two levels by driving with cross-cuts at 15 m intervals, with raises and cross-cuts	
4. Sampling	Core and sludge also for laboratory-scale beneficiation	Core and sludge underground channel examples, bulk samples for beneficiation test	Core and sludge also for laboratory tests	Core and sludge, channel sample, bulk samples etc.	Core and sludge for laboratory tests	Core and sludge channel samples, bulk samples	Core and sludge, core sample for laboratory tests	Core and sludge underground channel sampling, bulk sampling, etc.	

Table 8.6 Geological distribution of important lead-zinc deposits in India

State	District	Geological formation	Host rock	Important minerals
Rajasthan	Udaipur (Zawar)	Aravallis	Impure dolomitic limestone and dolomite	Sphalerite and galena
	Aguchha	Gneissic-complex?	Biotite gneiss?	Sphalerite and galena
Gujarat	Banaskantha (Ambamatamultimetal lodes)	Delhis	Talc and talc schists	Sphalerite, galena and chalco-pyrite
Orissa	Sundergarh (Sargipalle)	Gangpur series	Quartzwacke, quartzose clay and calcareous rocks	Galena, sphalerite and chalco-pyrite
Tamil Nadu	South Arcot (Mamandurmultimetal lodes)	Dharwar (?)	Contact of garnetiferous gneiss and biotite granite	Galena, sphalerite
Andhra Pradesh	Cuddapah (Zangamaraju-palle)	–	–	–

by the presence of ophiolites and greywackes in the sequence which contains the deposits. The absence of wall-rock alteration at Zawar has been observed by the proponents of the hydrothermal theory also. The general conformity of the ore bodies with the host rock, in parallelism with bedding planes, and the absence of any relationship with intrusive tend to support the view that mineralization in these deposits is syngenetic. Some recent work has indicated that some parts of the Zawar deposits show rhythmic lamination, load cast, slump structures and graded bedding which are all sedimentary structures. This suggests that at least a part of the Zawar deposit may be sedimentogenetic, later remobilized under favourable tectonic and hydrothermal conditions. The Sargipalle deposit has been considered as sedimentary on the basis of the following field evidence:

(1) Mineralization is confined to one stratigraphic horizon.
(2) Cross-cutting veins are absent,
(3) Variation in mineralization is correlatable with the variation in sedimentary facies,
(4) Metamorphism is isofacial in the ore and host rock,
(5) Intense shearing ore faulting along the ore zone is absent, and
(6) Wall-rock alteration is absent

Lead-zinc ore bodies generally show lenticular shapes, although some of the deposits may be banded. Thus, the lead-zinc prospects at Sargipalle occur in the form of bands and lodes. The Zawar deposits show clear strike continuity and considerable depth extension.

Prospecting and Exploration

The association of lead-zinc deposits with impure calcareous rocks gives a very good field guide in the prospecting operation. Old slag heaps, workings, etc. are equally helpful as in the case of copper. Geochemical and geophysical methods are also useful in the case of lead-zinc, just as in the case of copper. The lead-zinc deposits at Sargipalle were located and defined by self-potential survey.

Exploration

Exploration for lead-zinc deposits is summarized in Table 8.6.

(X) Pyrites

Two pyrites deposits of commercial value are known in India. One is located at Amjhor in Shahabad district of Bihar, and the other at Saladipura in Sikar district of Rajasthan. The deposits of Amjhor are of sedimentary origin while Saladipura deposit may be hydrothermal.

Prospecting and Exploration

Prospecting for pyrites is done more or less in the same way as copper, lead, zinc, etc. Geophysical prospecting (gravity and magnetic methods) gives very good response and may be used for locating and defining ore bodies.

The guides to exploration for vein-type deposits of pyrites have been given in Table 8.7. In the bedded type, one may have adopted the norms given in Table 8.7.

(XI) Gold

Five major types of gold deposits are recognized. They are

(1) Magmatic deposits
(2) Contact metasomatic deposits
(3) Hydrothermal deposits
    (a) Replacement
        (i) Massive
        (ii) Lode
    (b) Cavity filling
        (i) Fissure vein
        (ii) Stock works

Table 8.7 Guides to exploration of pyrites (bedded type)

	Method	Details
1.	Mapping	1: 8000
2.	Pitting	To be done near sca. Faces to expose thickness
3.	Drilling	500 m grid
4.	Exploratory Mining	Aditing

Table 8.8 Geological distribution of important gold deposits in India

State	District	Geological formation	Host rock	Important minerals
Karnataka	Kolar (Kolar Gold Fields)	Dharwar	Quartz vein in hornblende chlorite and mica schist (metavolcanics)	Sulphide
Andhra Pradesh	Raichur (Hutti)	Dharwar	Quartz vein in metavolcanics	Sulphide and native form
	Ananthapur (Ramagiri)	Dharwar	Quartz veins in sercecite, chlorite phyllite	Native
Kerala	Kozhikode (Wyanad)	Charnockites (Archaen)	Quartz vein in biotite granulite and hornblende granulite	Sulphide

(iii) Saddle reefs
(iv) Breccias

(4) Mechanical concentration
(5) Placer deposits

Of these, the only type recognized in India is the hydrothermal type. The geology and distribution of some important Indian gold deposits are given in Table 8.8.

Gold occurrences are reported from practically every state in India, although only a few of them are of economic importance. Excepting the occurrence of gold in charnockites in Wayanad (Kozhikode), all deposits are found in suites of metavolcanics in Dharwarian greenschists.

All Indian deposits are considered to be of hydrothermal origin. The exact source of hydrothermal solution is not clear. In Kolar, the mineralization has been genetically correlated with a suit of rocks known as Champion Gneiss, although this view is not university accepted.

In Ramagiri gold field, the host rocks of the gold deposits are andesitic lava flows. It is considered that the lava flows contained gold in a disseminated form originally. During metamorphism, the lava flows were reconstituted and mineralizing solutions containing gold emanated from them, and deposits were emplaced in quartz reefs. This hypothesis is more akin to the "Soutch Bed Concept" for sulphide mineralization rather than to the hydrothermal concept62. The "Source Bed Concept" supports the view that sulphide mineralization takes place syngenetically in sedimentary basins in particular horizons and they subsequently migrate to their present host rock during a rise in temperature (metamorphism?) of the enclosing rocks. Gold-bearing quartz veins occur within a suite of greenstones. In most cases, the veins remain parallel to the schistose host rocks. Cross-cutting veins are also seen. Sulphide-bearing lodes contain, in addition to gold, minerals like pyrite, pyrrhotite, galena, etc. The lodes generally form parallel fracture-filled vein systems. These vein lode systems show great strike continuity, narrow outcrop widths and conspicuous depth persistence as in the case of the Kolar Gold Fields where mineralization has been proved beyond a depth of 3000 m. Most ore bodies are characterized by steep dips. Dextral and sinistral cross-folds have affected the lodes and are considered to have influenced gold mineralization along with en echelon drag folds. Pinching and swelling along strike and dip are quite common in some of the reefs.

Prospecting and Exploration
Since gold mineralization is generally associated with acidic meta-volcanics of greenstone belts, the regional target areas should be selected in such areas.

Gold has been mined since ancient times in India, and it is usual to find old workings in many areas. These may include old pits, tailing dumps, slag heaps, etc. Kolar Gold Fields were located on the basis of old workings present all over the area. Trenching across old workings shows a compact clayey fill containing mined rock fragments and pieces of charcoal which were used presumably for "fire setting" which was an ancient gold mining practice. It is known that in Kolar the ancient mining was so selective and systematic that outcrops are traceable only at the bottom of large slumped wants heaps of gold workings.

In favourable target areas, pan washing for gold is an excellent prospecting tool. Gold placers in the form of specks, flakes and pinheads are usually found in sediments of stream running through the auriferous reef terrain. Systematic pan washing at intervals of 50 m, from the confluence to the watershed of several streams gave promising results in a gold prospecting operating in the Chandil area of Singhbhum district.

Since gold has a high specific gravity, even flakes and specks tend to concentrate at the bottom of the gravel bed. Hence, panning alone may not give conclusive results in many cases. Pitting to reach the bottom of the gravel bed is as equally important as pan washing. Random scout drilling may also be successful in reaching the bottom of the gravel bed.

Geochemical and geophysical methods are not very effective except where gold occurs along with other sulphides, like copper, lead and zinc. Many gold-bearing quartz reefs show the presence of scheelite. Since scheelite is fluorescent, and ultraviolet lamp survey in the suspected reefs may reveal the presence of scheelite which may lead to the location of gold.

8.2.3 Mineral Deposits Associated with Acidic Igneous Rocks

The major rock types in this group are granites and pegmatites. The associated mineral deposits are of tin, tungsten and mica. Various important minor minerals are quartz, feldspar, beryl, tourmaline, sapphire, lithium, rubidium, cesium, etc. Of these, tin, tungsten and mica are important in the Indian context.

(XII) **Tin-Cassiterite**

Three types of cassiterite deposits are recognized. They are

(1) Hydrothermal deposits

   (a) Stock works
   (b) Fissure veins, and
   (c) Disseminated replacements

(2) Pegmatitic deposits
(3) Placer deposits

Of these, the last two types are known in India, although so far there has been only one major potentially economic find the tin fields of Bastar district in Chhatishgarh. Here, the primary source of cassiterite is a zoned pegmatite. Some fossil placers are also considered promising.

The lepido-litepegmatites which show cassiterite mineralization I Bastar have been emplaced in metabasic sills which have intruded the Bengal group of metasediments. Though the Palim granite located closely is to have given rise to the gegmatites. Cassiterite occurs as discrete crystals in a disseminated form within the pegmatites along with rich veins of solid cassiterite ore. Mineralization took place by replacement. The pegmatites have intruded along the contact of metasediments with the metabasics. The joint planes of serecite-quartzites have also helped the pegmatitic intrusion. Pegmatite bodies containing cassiterite occur as irregular lensoid concentrations within shear zones. They have highly irregular shapes. Placer deposits of cassiterite are derived from these pegmatites and show highly irregular concentrations.

Prospecting and Exploration

Tin-cassiterite deposits of pegmatitic and placer types usually occur together or in close proximity as is seen in the case of Bastar tin field. Most of the tin deposits are confined to Precambrian terrain where large granitic plutons have been emplaced. Regional targets, therefore, should be selected from Precambrian areas showing granitic intrusions. Large target areas should be subjected to geochemical surveys. The methods generally in use are stream sediment and suboutcrop sampling. Stream sediment sampling was sued by Geological Survey of India intensively in their investigations for tin in Bastar district of Chhatishgarh and was mainly responsible for the initial discovery of the deposit. In a prospect in Hazaribagh district, the target, granitic gneiss, was subjected to suboutcrop sampling which helped in confirming that the mineralization was too spotty to be economical.

Any promising area should be selected for photogeological studies (1:30,000 or larger scale photography) to locate pegmatite occurrences and also to build a geomorphological map of the area. The geomorphological approach is necessary because the level of erosion of the terrain has a profound influence on the formation of placers. Features which need to be specially studied are alluvial cones, fans, river terraces, meander bends, river confluences, abandoned river channels, etc. Various potential targets should be marked on the photographs and test pitting carried out in these targets. Bulk samples should be collected and the samples concentrated by jigging or pan washing to see whether cassiterite pebbles and fragments are present. The presence of cassiterite in any quantity warrants detailed exploration of the

Table 8.9 General guidelines for exploration work for tin, tungsten and gold

	Method	Regional exploration stage	Intensive exploration stage	Remarks
1	Mapping	1:2000–1:5000	1:2000–1:5000	Most gold deposits are however explored by the same methods as in sulphides. See guides for sulphides in Table 8.13
2	Surface drilling	200–500 m section interval in two levels, 30–60 m vertically apart	100–200 m section interval in 2–3 levels, 30–60–90 m vertical interval to trace and intersect mineralized zones	
3	Underground drilling	Not recommended	As and where necessary	
4	Trenching	200–300 m section intervals to trace old working	Not recommended	
5	Exploratory mining	Not recommended	3 or more levels over the entire/part strike length of ore body at 30 m level interval and winzes along suitable intervals	
6	Sampling	Core and sludge, blast and channel	Core and sludge, blast and channel, bulk sample from underground developments for beneficiation test	

target. Geophysical prospecting is useful in locating buried river channels and certain types of riverine accumulations Trenches and pits may become necessary during outcrop mapping (Table 8.9).

(XIII) Tungsten–Wolframite–Scheelite

Four types of tungsten deposits are known

(1) Pegmatitic deposits
(2) Contact metasomatic deposits
(3) Hydrothermal deposits

   (a) Replacement deposits
   (b) Fissure vein

(4) Placers

Prospecting and Exploration

Tungsten deposits of primary origin as well as their secondary derivatives tend to occur in the same terrain. Prospecting should aim at locating both types of deposits.

Table 8.10 Geological distribution of tungsten deposits in India

State	District	Geological formation	Host rock	Important mineral
Rajasthan	Nagaur (degana)	Archaean	Granites and phyllites	Wolframite
West Bengal	Bankura	Dhonjori (Dharwars)	Quartz veins and alluvium	Wolframite
Maharashtra	Nagpur	Archaeans	Veins of quartz interbedded with schists	Wolframite
Karnataka	Kolar (Kolar Gold Fields)	Dharwars	With gold-bearing quartz veins	Scheelite

As mineralization occurs close to intrusive granitic bodies, the regional targets should be chosen from area with a known history of granitization. Geochemical sampling is useful in locating areas of mineralization. Both soil and stream sediments need be sampled. Geophysical methods (magnetic and electrical resistivity) have proved ineffective in the case of wolframite. Areas of scheelite mineralization can be recognized by ultraviolet kamp surveys. Scheelite, being fluorescent, shows up in ultraviolet light. Photogeological studies along the same lines suggested for tin prospecting may be useful in the case of tungsten deposits also. In India, the known deposits are either contact metasomatic, hydrothermal or placers (Table 8.10).

(XIV) Mica

Mica occurs in book form in pegmatites and is the only type of deposit of economic value. Several types of mica are known but in India only muscovite and phlogopire are of economic value. Of these, presently only muscovite is being mined. The various types of mica are listed below

(1) Muscovite (potassium mica)
(2) Paragonite (sodium mica)
(3) Lepidolite (lithium mica)
(4) Lepidomelane (iron mica)
(5) Phlogopite (magnesium mica)
(6) Zinwaldite (lithium-iron mica)
(7) Biotite (magnesium-iron mica)

Mica-bearing pegmatites may be of three types viz., (i) fracture filling, (ii) replacement bodies and (iii) zoned pegmatites. Of these, the last two are of economic value.

Prospecting and Exploration

Prospecting for mica consists of two tasks. One is to locate pegmatite bodies and the other to locate mica deposits within the pegmatites. Field guides for locating mica deposits have been given in Table 8.11. Regional targets are chosen from areas where mica schists have been intruded into by granitic plutons. The presence of lit-par-lit injections within the country rock is a negative indicator and such areas should be rejected.

Table 8.11 Guides to exploration of mica

	Method	Details
1	Mapping (surface) (Underground)	1:1000 1:500 1:100–1:50
2	Trenching	30 m along strike
3	Exploratory mining	8 m interval levels, winzes and raises, cross-cuts at close intervals
4	Drilling	Extension rods fitted to jackhammer and other methods in between cross-cuts and advancing faces

In locating areas of pegmatite intrusion, aerial photographs are very useful. Even concealed bodies can be recognized by studying the pattern of exposed pegmatites, by photogeological studies.

After locating pegmatite bodies, it is necessary to check whether they are mica-bearing or not. The presence of various indicator minerals, zoning, etc. generally indicates the likelihood of mica deposits. Laboratory-scale studies to recognize the mineral assemblage also help in recognizing potential mica-bearing pegmatites.

8.2.4 Mineral Deposits Associated with Sedimentary Evaporate Rocks

The sedimentary evaporate environment encourages many types of deposition, photosynthetic accumulations and evaporates. One common factor is the presence of water in circulation or stagnant water or both in all these processes. The most important mineral deposits in this group are limestone, dolomite, gypsum, phosphorite and rock salt.

(XV) **Limestone and Dolomite**

Limestone and dolomite are formed from solutions carrying calcium and magnesium carbonates. Minerals and rocks containing carbonates in major or minor quantities are the original source of all limestone and dolomites. Limestone may form by three processes, viz., inorganic, organic and mechanical.

Carbonates are precipitated when carbon dioxide escapes from the solution. The amount of carbon dioxide contained in sea water is dependent upon the temperature of water and the saturation of carbon dioxide in the atmosphere. When the equilibrium between these three is disturbed, carbon dioxide escapes and carbonates are precipitated in the form of limestones, as warm sea water is usually saturated with calcium carbonates. Inorganic limestones are formed this way.

In the organic process, action of algae, bacteria, corals and other micro-organisms are responsible for calcium carbonate precipitation. In the mechanical process, shell or coral matter gets deposited and cemented to form limestones. Dolomite is a carbonate of calcium and magnesium and its origin is still an unsettled issue. However, most of the dolomites are thought to be of replacement origin.

Prospecting and Exploration

In selecting broad target areas of limestone and dolomite, photogeological mapping is of great help. The presence of sinkholes and karsts in a generally flat and subdued topography indicates the presence of limestone. Limestone terrains also encourage subsurface drainage due to the presence of solution cavities. These features are very easily recognized in aerial photographs. The ideal photographic scale for this type of work is 1:50,000.

Soils which develop on limestone tend to contain a high level of moisture which gives a dark tone to the soil can be easily spotted in the aerial photographs and is another indication of the presence of limestone.

Some chip sampling and rapid analysis for CaO, MgO, etc. would help in confirming the presence of limestone and dolomite (Table 8.12).

(XVI) Gypsum

Four major types of gypsum deposits are recognized in India. They are

(i) Bedded evaporate sedimentary type,
(ii) Crystalline gypsum in veins or disseminations in clays and shales,
(iii) As grains of gypsum in semiporous aggregates in desert salt pans, and
(iv) As boulder disseminations in soil.

Prospecting and Exploration

Gypsum usually occurs along with limestone and clay beds. The initial target areas may be those which are known to contain limestone. Since gypsum resembles limestone to a great extent, chemical tests are essential to identify the two separately, for which chip sampling may be done.

Gypsum also occurs in drying lakebeds which receive a periodic supply of sea water. Areas where such lakes are known are good targets in the search for gypsum. Aerial photographs may be used for recognizing such targets.

The nodular deposits tend to occur in the subsoil and its presence is indicated by boulders of gypsum within the soil.

Vein-type deposits are difficult to locate. Areas having limestones and pyrites in close proximity generally contain vein-type gypsum deposits. Large-scale systematic mapping accompanied by chip sampling would appear to be the best method to locate such deposits.

(XVII) Phosphorite

Phosphorite deposits are invariably chemico-sedimentary in nature and are found only in the sedimentary formations. Phosphorite occurs in seven forms. There are

(1) granular phosphorite,
(2) pelletal phosphorite,

Table 8.12 Exploration guidelines for limestone

Method	Simple type		Complicated type		Highly complicated type		Remarks
	Regional exploration stage	Intensive exploration stage	Regional exploration stage	Intensive exploration stage	Regional exploration stage	Intensive exploration stage	
1. Mapping	1:2000–1:5000	1:1000–1:2000	1:1000	1:1000	1:1000–1:2000	1:1000	
2. Drilling	400–600 m interval	200–300 m grid	300–400 m section interval, 1–2 boreholes in each section	150–200 m section interval	100–200 m section interval 1–2 boreholes in each section	50–100 m section interval	
3. Pitting	2–4 Nos. for every sq. km	2–4 Nos. for every sq. km	Not recommended				
4. Trenching	Not Recommended		150–200 m section interval	75–100 m section interval	100–200 m section interval	50–100 m section interval	
5. Sampling	Core, blast and channel	Core blast and channel sample	–				

(3) banded phosphorite,
(4) phosphatic nodules,
(5) algal phosphorite,
(6) brecciated or fragmental phosphorite, and
(7) fossilliferousphosphorite.

Prospecting and Exploration
Areas of miogeosynclinal deposition are chosen as the first broad target area for prospecting. Economic deposits are associated with limestone, calcareous cherts, black shales, etc. and areas containing such rocks are separated out for detailed examination. Most phosphorite beds contain radioactive minerals. Therefore, a rapid survey by a Geiger–Muller counter or scintillation equipment would help in locating promising areas of mineralization. This is followed by a millimetre by millimetre traverse of favourable outcrops in which different beds are recognized by acid test and scintillation counts. Phosphorite does not react to acid but activates the scintillation instrument. This is a positive criteria on for the recognition of phosphorite in the field. Chip samples are collected from the suspected phosphorite bands and subjected to chemical analysis by Shapiro's method. The method described above was used in a prospecting operation for phosphorite in Turkey which resulted in important discoveries.

Exploration guidelines for phosphorites are shown in Table 8.13.

8.2.5 Mineral Deposits Associated with Metamorphic Rocks

In this group, mineral deposits like kyanite, sillimanite, graphite, talc, etc. are the most important, since these deposits are undeniably metamorphic in origin. Iron and manganese ores are included in this group, although their origin cannot be attributed to metamorphism. Their host rocks are general meta-sedimentary and hence these two are included in this group.

(XVIII) **Iron Ore**

Six types of iron ore deposits are recognized in India. These are

(1) deposits associated with the banded ferruginous formations of Precambrian Age. These include both hematitic and magnetitic types,
(2) sedimentary iron ores of sideritic or limonitic composition,
(3) lateritic ores derived from the subaerial alteration of the iron ore bearing rocks, such as gneisses, schists, basic lavas, etc.
(4) the apatite-magnetite rocks of Singhbhum copper belt,
(5) the titaniferous and vanadiferous magnetites of southeast Singhbhum and Mayurbhanj, and
(6) fault and fissure fillings of hematite.

Table 8.13 General guidelines for exploration of phosphorite

Method	Flat or low dipping uniform bodies		Complicated deposits with variable thickness etc.	
	Regional exploration state	Intensive exploration stage	Regional exploration stage	Intensive exploration stage
1. Mapping	1:2000–1:5000	1:1000	1:1000	1:1000
2. Drilling	200–300 m grid	100–200 m grid	200–300 m section interval, 1–2 boreholes per line to intersect 50/100 m levels	100–150 m section interval intersection at 2–4 levels
3. Pitting	4–10 Nos. per sq. km	4–6 Nos. per sq. km	Not recommended	
4. Trenching	Not recommended		200–300 m section interval	100–150 m section lines
5. Exploratory mining	Not recommended			Two levels development by drives/winzes, cross-cuts at approximately 30 m interval
6. Sampling		Core and sludge, blast and channel samples, bulk samples for beneficiation		

All the major Indian iron ore deposits belong to the first group and are associated with banded iron ore formations of Precambrian age. Hematite ore is under extensive exploration and mining. Magnetite ores have started receiving attention only recently. Some of the major iron ore deposits of India are listed in Table 8.14.

The other occurrences reported in various geological formations are not of any economic importance.

As mentioned earlier, all the iron ore deposits of India belong to the Iron ore series and their equivalents of the Dharwar group of meta-volcanics and metasediments. The immediate host rock is the banded ferruginous quartzite. These banded formations were probably formed as chemical precipitates in partially enclosed sedimentary basins of the back water type, in long periods of geological quiescence. It has not been possible to identify the original source of the iron and silica minerals which were precipitated in such basins.

The deposits are considered to have been derived from the enrichment of the Banded Iron Stone Formations by a process of progressive removal of silica. Controversies, however, remain about the development of massive ores and blue dust. One view is that the massive ores are the resultants of direct precipitation. Blue dust is thought to be a product of the action of circulating waters which leach away the silica from the massive ores.

The iron ore formations within the Dharwars have been folded repeatedly and are preserved in narrow elongated patches of schists belts, a few of them forming synclinoriums like the Sandur synclinorium in Bellary district. It has been observed that a very large number of deposits in Karnataka and Goa show typical structural

Table 8.14 Geological distribution of important iron ore deposits in India

State	District	Geological formation	Host rock
Andhra Pradesh	Anantapur	Dharwara	Banded ferruginous quartzite
Jharkhand	Singhbum	Iron ore series (Dharwareq)	Banded hematite quartzite
Madhya Pradesh	Baster (Bailadila) Durg	Bailadila Iron Ore Series (Dharwareq)	Banded hematite quartzite
Maharashtra	Ratnagiri Chanda	Dharwar Dharwar	Not clear Not known
Goa		Dharwar	Banded ferruginous quartzite
Odisha	Keonjhar (Malangtoli) Sundargarh Mayurbhanj	Iron Ore Series (Dharwar equivalent) – –	Banded hematite quartzite – –
Rajasthan	Jaipur	–	–
Karnataka	Bellary (Sandur) Bijapur Chitradurga North Kanara Tumkur	Dharwar	Banded ferruginous quartzite
	Chikmagalur (Kudremukh), (Bababandan, etc.)	Dharwar	Banded ferruginous quartzite
	Shimoga	Dharwar	Banded ferruginous quartzite
Tamil Nadu	Salem (Kanjamalsi, etc.)	Dharwar	Banded magnetite quartzite

features. The major deposits in these two areas appear to have been preserved in tightly folded regional synclinal structures. The individual iron ore deposits are preserved in local structures, predominantly synclinal. However, cappings and narrow bands which show no obvious structures are also quite common.

Most hematitic ore bodies can be grouped into two categories on the basis of their shape and mode of occurrence as capping and reef types. The major difference between the two is the dip, which will be very steep in the case of reefs and rather shallow in the case of cappings; outcrop widths may be narrow in the case of reefs and broad in the case of cappings. These differences may not be very sharp in many cases as both reef and capping types may have a common origin and tectonic background.

Prospecting and Exploration

Hematitic and magnetitic iron ore bodies are generally either confined to the surface or are very close to the surface. Their location is not likely to create any problem. Field guides which may help in the location of such ore bodies are given in Table 8.14.

Iron ore bodies may be found either to be exposed directly ore covered by laterite, soil or other detrial material. The directly exposed ore bodies can be recognized by the hard massive ore outcrops of iron ore. Those under soil, laterite, etc. can be recognized by the presence of float ore or recemented or boulders embedded within laterite or soil.

When ore bodies are not exposed and there are no obvious floats or other indications, it may become necessary to rely on a combination of indirect criteria to locate favourable targets. These may consist of (i) the presence of BHQ/BHJ, etc. (ii) topographic prominences, (iii) laterite with ore pieces, and (iv) structural ridges. On concealed outcrops, scout pitting and trenching may be done to confirm the presence of ore. Guidelines for exploration or iron ore are summarized in Table 8.15.

(XIX) Manganese Ore

Three types of manganese deposits are recognized in India. They are

(1) deposits associated with Precambiran metamorphic rocks of Gondite type,
(2) deposits associated with Precambrian metamorphic rocks of Kodurite type, and
(3) lateritic deposits in surficial concentrations with mineralization extending into the underlying rocks.

All the three major deposits belong to the Dharwar group of rocks and their equivalents. The lateritic deposits are almost invariably found in association with the Iron Ore Series of rocks and their equivalents and occur mostly in the stratigraphic horizon just below the Banded Iron Formation. They are exploited on a small scale locally. The important manganese deposits of India are listed in Table 8.16.

Just as in the case of iron ore, the association of manganese with the Dharwar rocks and their equivalents is unmistakable. Besides, there would appear to be a definite, broad lithological similarity between the host rocks, most deposits being confined to pure argillaceous and arenaceous rocks and their metamorphic products. Carbonaceous host rocks are seen in the Madhya Pradesh–Maharashtra belt and in some parts of the Tumkur-Chitradurga belts in Karnataka.

The economic manganese ore deposits appear to have close genetic association with various types of low-grade manganiferous sediments. The deposits of Madhya Pradesh and Maharashtra are thought to have been derived from spessartite quartz rocks or gondites and similar other rocks. The ore bodies grade into gondites as they are traced along the strike. These spessartite quartz rocks are all thought to be metamorphic products of low-grade manganiferous sediments. One of the products in this metamorphism was probably braunite bearing manganese ore bodies.

Table 8.15 General guidelines for exploration work for iron ore (hematite) (excluding purely lateritic deposits)

Method	Capping type deposits		Reef type with appreciable dip		Remarks
	Regional	Intensive	Regional	Intensive	
1. Mapping (a) Underground mapping for adits	1:2000–1:5000	1:1000–1:2000 1:200	1:2000–1:5000	1:1000–1:2000 1:1000–1:200	To map lithology and boundaries, soil-ore to waste contact ore types and their contact, structural features, etc. Sludge collection is important wherever core loss occurs. Dry drilling useful in soft zones.
2. Drilling	100–500 m section intervals 1–2 boreholes each section	50–150 m section interval, 2–4 boreholes in each section to outline bottom of the ore	100–300 m section interval along two levels	100–150 m section interval down to 90 m depth	
3. Pitting	2–4 nos.	Deep pits down to 15 m depth, 2–4 nos. on each section line to determine lump: fine ratio etc.	Up to 3 deep pits	1–2 nos. in alternate for every third section	
4. Aditing	Not recommended	Cross-cutting adit intersecting ore body 2–3 nos. along representative sections with 30–50 m back	Not recommended	2–6 adits at different levels and at 300–500 m, lateral interval	
5. Sampling	Core and sludge, bulk sample from pit	Cores, sludges, bulk from pits and adits for every 1–2 m, interval for grade and size classification	Core and sludge, bulk sample for every 1–2 m depth from pit for size and grade classification	Core and sludge, bulk sample for every 2 m depth from pits and adits for grade and size classification	

Table 8.16 Geological distribution of manganese ore deposits in India

State	District	Geological formation	Host rock	Important minerals
Andhra Pradesh	Adilabad	Cuddapah	Limestone or shales with inter-banded chert	Psilomelane, pyrolusite and braunite
	Srikakulam Vishakhapatnam	Dharwar	Garnet granulite rocks (Argillaceous facies) Garnetiferous quartzite, (Arenaceousfacies)	Psilomelane, pyrolusite and braunite
Jharkhand	Singhbhum	Iron ore series	Laterities, phyllitesshales and cherty quartzite	Pyrolusite, psilomelane
Goa	–	Dharwar	Laterite, phyllite and quartzite	Pyrolusite and psilomelane
Karnataka	North Kanara Chitradurga Shimoga Tumkur Bellary	Dharwar	Laterite, phyllite and cherty quartzite and crystalline limestone	Pyrolusite, psilomelane, manganite
Madhya Pradesh	Balaghat Chindwara Jhabua	Sausar series	High grade metamorphic products of argillaceous, carbonaceous and arenaceous sediments	Braunite, pyrolustie and psilomelane
Maharashtra	Bhandara Nagpur			
	Ratnagiri	Dharwar	Phyllites and cherty quartzites	Pyrolusite, psilomelane
Odisha	Bolangir Koraput Keonjhar Sundergarh	Dharwar iron ore series	Khondalities and kodurites, laterites, pyllites, shales and cherty quartzites	Pyrolusite, psilomelane
Rajasthan	Banswara	Aravallis	Phyllites and quartzites	Braunite, pyrolusite, psilomelane
Gujarat	Baroda	Aravallis	Phyllites and quartzites	Pyrolusite, psilomelane, braunite

Alteration of the braunite is thought to have produced psilomelane, pyrolusite, etc. The ore bodies are bedded and occur in conformity with the host rock.

The deposits of Srikakulam area referred to as the Kodurite type are also unmistakably associated with two types of metasediments, viz. (i) a garnet granulite, and (ii) a garnetiferous quartzite. Thus, deposits of Srikakulam of Andhra

Pradesh, Koraput of Orissa and other deposits of Madhya Pradesh, Maharashtra are related to the process of metamorphism.

In the lateritic type, there are two types of deposits. One occurring within laterite cappings and the other occurring in the underlying phyllites, quartzites and shales. Of these, the deposits within the lateritic cappings are of lateritic origin but derived from the underlying rocks. The deposits within the quartzites, phyllites and shales may be of sedimentary origin as indicated by the conformity of the ore bodies with the host rocks and the presence of continuous planar features within the deposits and the host rocks.

All manganese deposits including those associated with laterites have been influenced and controlled by structure. The major structural features recognized in the Madhya Pradesh–Maharashtra belt are isoclinals folds as shown by the ore body. Structural control is prominent in the Srikakulam deposits also where the ore shows preferential concentration along the noses of drag folds.

In the lateritic deposits, the laterite cappings show deposits totally devoid of any structure. But the underlying phyllites and quartzites have preserved ore bodies in regional scale isoclinals synclines and recumbent folds as at Sandur. Here, cross-folds have played a part in bringing the important ore bodies to the surface.

Prospecting and Exploration

Manganese ore bodies occur in a variety of forms and in varying depths. Purely surficial deposits are confined to the lateritic types, although here also some deposits are known to persist for some depth. The major surface indication of the presence of manganese ore is the presence of float ore. The prospecting history of many major deposits indicates that the first discovery was made on the basis of float ore. A comprehensive list of field guides for locating manganese deposits is given in Table 8.17.

Where outcrops are not directly visible but are inferred, geophysical methods may help in locating the ore bodies. Electrical methods are particularly useful in locating, the ore bodies in deposits of the gondite type. These methods may, however, prove useless in the case of lateritic type of deposits where the country rock has a high iron content.

Pitting and trenching would be useful in locating outcrops. The presence of recemented ore pieces in narrow linear lateritic humps is considered very favourable in locating deposits underneath as in the Sandur Manganese field.

(XX) Kyanite and Sillimanite

Commercial deposits of kyanite occur in two forms, viz. along with metamorphic rocks, and as a product of residual concentration as flat ore, etc.

Prospecting and Exploration

In prospecting for kyanite and sillimanite, broad targets are selected from areas containing aluminous sediments which have undergone high-grade metamorphism. Mineralization can be easily established by tracing of float ore and systematic shallow pitting.

Exploration guidelines for kyanite and silliminate are summarized in Table 8.18.

Table 8.17 General guidelines for exploration work for manganese ore

Method	Sample from strata-bound deposits		Banded deposits with complicated structure		Lateritic bodies		Remarks
	Regional exploration state	Intensive exploration stage	Regional exploration stage	Intensive exploration stage	Regional exploration stage	Intensive exploration stage	
1. Mapping	1:2000–1:5000	1:1000–1:2000	1:1000–1:2000	1:1000–1:2000	The main method of exploring lateriticmanganese bodies is by pitting and trenching, if possible by shallow drilling in order to delineate individual bodies including blind ore bodies		
2. Underground	Not recommended	1:2000	Not recommended	1:1000–1:2000			
3. Pitting/trenching	As necessary		50–100 m. Trenching section interval wherever feasible				
4. Drilling	200–300 m section interval at 2 levels with 60 m vertical interval	100–200 m section interval intersection at 3–4 levels	50–100 m section interval at 2–3 levels with 30–60 m vertical intervals	25–50 m section interval at 3–4 vertical interval			
5. Exploratory mining	Not recommended	Two levels across entire strike length	Not recommended	2–3 levels 20–30 m vertical interval where necessary			
6. Sampling	Core and sludge channel and blast pit/core sample for laboratory study	Core and sludge channel and blast for pilot plant beneficiation studied	Core and sludge channel and blast etc. for laboratory-scale beneficiation	Core and sludge channel and blast for pilot plant beneficiation test			

Table 8.18 Guides to exploration of kyanite

	Method	Details
1	Mapping	1:1000–1:5000
2	Pitting/trenching	30 m interval, trench width varying from 1 to 2 m or more
3	Sampling	From pits and trenches for analysis and recovery tests

Table 8.19 Guides for exploration of graphite, talc–soapstone pyrophyllitediaspore

	Method of exploration	Details
1	Mapping/underground	1:1000–1:4000 1:2000–1:5000
2	Pitting	May be given at suitable intervals
3	Trenching	Cross trenching, 30–50 m a part may be given
4	Drilling	50 m interval with a staggered pattern depending upon the mineralization
5	Sampling	Trench sampling, core sampling and channel sampling

(XXI) Graphite

Graphite occurs in commercial quantities in high-grade metamorphic rocks. The geology and distribution of the major graphite deposits of India are shown in Table 8.19.

Graphite is a product of high-grade metamorphism of argillaceous carbonaceous sediments. Deposits generally occur as veins, lenses and dissemination in the metamorphic rocks.

Structural controls have been recognized in the graphite mineralization of Orissa. Here, mineralization is noticed at the contact of khondalites and granites. Shear zones developed in the contact zones show migmatization which has helped in localizing the deposits. Deposits occurring in pegmatite are more pure in quality. The ore bodies show general concordance with the planner features of the host rock.

Prospecting and Exploration

Areas of high-grade metamorphism particularly of argillaceous carbonaceous sediments are very good targets. Geophysical prospecting gives excellent anomalies in graphite. Electrical methods (SP) were successful in locating and delineating graphite ore bodies in Ernakulam district of Kerala. The anomalies were subsequently proved by core drilling.

Exploration guides for graphite are similar to those for talc, soapstone, etc. are given in Table 8.19.

Prospecting and Exploration

Metamorphic terrain which contains magnesium carbonate rocks forms the best target area for talc and soapstone. For pyrophyllite and diaspore, rapid reconnaissance, outcrop mapping, chip sampling, etc. help in locating individual deposits.

The typical features which help in identifying talc and pyrophyllite are their softness and powdery glossiness. Deposits tend to occur in large clusters of lenses.

Exploration guidelines for talc soapstone and pyrophyllites are summarized in Table 8.19 based on the experience of Indian Bureau of Mine.

8.2.6 Mineral Deposits Associated with Residual Formations

A large number of mineral deposits occur as residual enrichments. The most important ones are iron ore, manganese ore, nickel, bauxite and clay. Of these, however, most of the deposits were formed originally in other environments and residual action has only augmented them. Typical residual deposits are bauxite and clay.

(XXII) **Bauxite**

Bauxite generally occurs in association with laterite and is of residual origin. Most Indian deposits are seen within caps of laterite occurring on a variety of rock types. Some deposits are known to occur in soil and some in sediments, but all are clearly derived from original lateritic sources nearby.

Prospecting and Exploration

Since bauxite is genetically associated with laterite, any lateritic terrain is a potential target for bauxite prospecting. Regional small-scale aerial photographs are ideally suited for locating potential bauxite-bearing laterites. Bauxite discourages the growth of luxuriant vegetation. In the aerial photographs, such areas stand out conspicuously and can be easily identified.

Scarp retreat is a common phenomenon of bauxite-bearing plateau country. In this, the edge of the plateau recedes towards the center of the plateau by progressive weathering. Such areas show the outcrops of bauxite clearly. However, the absence of scarp retreat is not a negative indicator.

In India, most of the important bauxite deposits occur on plateaus which rise to 1000 m above the mean sea level. This criterion can be used for isolating the potential target areas.

Bauxite does not differ much from laterite in its physical appearance and it may be difficult to distinguish them without chemical analysis. Therefore, at the initial stages, rapid chip and channel sampling and chemical analysis of promising outcrop areas is very essential. Samples which show a loss on ignition of less than 20% are considered very poor. Deposits which show such poor values can be rejected on this basis alone. Geophysical and geochemical methods have only limited value in bauxite prospecting. Exploration procedures for bauxite are summarized in Table 8.20.

Table 8.20 General guidelines for exploration work for bauxite (also applicable for residual clays, nickel of lateritic origin and other lateritic deposits, manganese and iron ores)

Method	Bedded extensive deposits with uniform grade		Lenticular extensive deposits with variable grade		Lenticular extensive deposits with variable grade		Remarks
	Regional exploration stage	Intensive exploration stage	Regional exploration stage	Intensive exploration stage	Regional exploration stage	Intensive exploration stage	
1. Mapping	1:2000–1:5000	1:1000	1:2000–1:5000	1:1000	1:2000–1:5000	1:1000	To delineate contacts between bauxite and laterite, bauxite, clay and formation of host rock
2. Drilling	200 m interval, a few pilot holes to touch basement 50 m grid in test strip	100 m interval down to top of lithomarge	100 m intervals down to top of lithomarge, a few holes to touch basement, 25 m grid in test strip	50 m centres down to top of lithomarge	2–4 nos. down to basement	1:1000	
3. Pitting	4–10 nos. down to top of lithomarge for 1000 m strike length and 100–300 m width	100 m centres half way between boreholes	100 m centres half way between bore-holes down to top of lithomarge	50 m centres half way between boreholes down to top of lithomarge	50 m centres, a few pilot pits down to lithomare, test strip study on 25 m centres	25 m center	
4. Trenching	Not recommended	Not recommended	Not recommended	Not recommended	Not recommended	Long trenches to establish the average recoverability of bauxite	
5. Sampling	Channel sample at 20 m interval along scarp exposure, channel and bulk samples from pits laboratory-scale beneficiation studies	Core and sludge bulk and channel from pits, lump and fine to be analysed	Core and sludge, channel sample at 10 m intervals along scarp exposures. Bulk and channel from pits	Core and sludge and bulk sample from pits, blastwise analysis of lump and fine separately	Bulk and channel samples from pits	Bulk and channel samples from pits	Scrap cutting is an important sampling practice

(XXIII) Clays

Various types of clays are known of which the most important are china clay and fireclay. The other clays are ballclay, bentonite, pottery clay, fuller's earth, diaspore-clay, etc. The exploration methods for china clay and fireclay, which are applicable to all other clays also, are described below:

(a) ***China clay***: China clay is a product of residual weathering of felspathic rocks in favourable climatic and geological conditions. The main clay mineral which is kaolin can form by the action of water of post-magmatic origin. Kaolinization may also occur as a result of the action of post-magmatic emanations or by a process of hydrolysis.

Prospecting and Exploration

Being a product of residual weathering similar to bauxite, prospecting for clay is essentially similar to that of bauxite. However, there are many important differences. In selecting large areas for initial prospecting of china clay, it is better to confine attention to areas which have a known history of residual weathering, and contain acidic rocks like granite and pegmatite in granite and pegmatite. The aerial photographs, clay and clayey soil give very dark tonnes because of high moisture supported generally by good growth of vegetation. Swamps, marshy lands, water-filled depressions, etc. may also indicate the presence of china clay. Rapid reconnaissance with random pits and auger drilling will help locate specific targets for detailed search.

China clay deposits usually occur near the surface and methods of exploration may consist of mapping, pitting, drilling, etc. which gives exploration guidelines, for bauxite. Further, the clay being soft, conventional wet core drilling may be of limited value, dry drilling, auger drilling or pitting being better suited.

(b) ***Fireclay***: Fireclay is generally associated with coal seams and is of sedimentary origin. Fireclay contains Al_2O_3 above 31% while good-quality fireclay may contain about 38% Al_2O_3. Fireclay occurs both above and below coal seams in this bedded sedimentary form, showing same dips of coal seams which are 10–20°.

Prospecting and Exploration

As most of the important clay deposits occur in association with the coal-bearing Gondwana formations like Barakar, Raniganj and Karharbari formations, the initial target areas can be chosen from these formations. Since fireclay beds occur above or below the coal seams, all methods valid for locating the seams are valid for the fireclays too. Coal seams may be directly visible if they occur in the escarpment faces as in Chirimiri coalfields, Madhya Pradesh. However, in most cases, indirect evidences will have to be depended upon. In areas where coal seams outcrop, there will be usually linear depression which retains high moisture. These depressions are used for paddy cultivation and are an excellent field guide for locating coal seams. The presence of fireclay can be confirmed by putting shallow pits towards edges of such depressions which coincide with the top or bottom of coal seams. In such sections, generally plastic fireclay may occur at the top followed by non-plastic clay. Black soil, carbonaceous particles, presence of coal matter, etc. are also good indications for the presence of coal seams.

However, the best areas for search would be opencast coal mines where fireclay beds, if present can be seen outcropping below the coal seams. A search through the old dumps of abandoned opencast mines can also give a clue to the presence of fireclay. An examination of exposed seam sections also helps in locating the fireclay beds.

Since fireclay can be mined economically only when it occurs within a depth of say 30 m (maximum) or so, it would be futile to look for it in deep underground coal mines.

Beds fireclay occur independent of coal seams within coal-bearing sequence. However, the methods of prospecting already discussed are valid.

8.2.7 Mineral Deposits Associated with Placer Formations

Important placer deposits known in India are monazite-bearing sands of Kerala, Tamil Nadu, Andhra Pradesh and alluvial tin deposits of Madhya Pradesh. Of these, the monazite-bearing sands contain, in addition to monazite, a number of other minerals like sillimanite, ziron, ilmenite, etc. These sands are essentially beach placers in which are concentrated the minerals from riverine sediments. All the other placers are of riverine alluvial origin.

Deposits of gypsum are known in certain sand dunes of Rajasthan, which should be considered as an aeolian placer.

8.2.8 Miscellaneous Mineral Deposits

Because of their diverse origins, and specific the following minerals which are of economic importance have been grouped together. No prospecting guides or exploration guides can be offered for these minerals. However, whenever one of them resembles a major mineral described earlier, its exploration sequence may be comparable. In this group, various semi-precious stones, silica sand, feldspars, vermiculite and quartz are included.

Further Readings

Arogyaswami RNP (1988) Course in mining geology, 3rd edn. Oxford IBH, p 695

Burn RG (1984) Exploration risk. CIM Bu 77(870):55–61

Guilbert JM, Park CF (1986) The geology of ore deposits. W.H. Freeman, San Francisco, p 985

Hustrulid WA, Bullock RL (eds) (2001) Underground mining methods engineering fundamentals and international case studies. Society for Mining. Metallurgy and Exploration. Inc., (SME), USA, p 718

Chapter 9
How to Arrive at Decision to Expend Operating Mines?

9.1 Exploration in Producing Mines

In the earlier chapters, we have discussed the location of prospects, prospecting and exploration of mineral deposits to reach at a conclusion to open a mine. Since to open a mine, all the basic data regarding the deposit is available, here we deal with the exploration needs of producing mines. It is necessary to translate available data into specific mining requirement. Each mine is designed to produce a certain quantity of ore each day. The quantity being mined now needs to have specific average grade and size, fixed according to utilization of the ore. The basic problem is to locate the ore of specific quality from the benches or underground mining faces. Mining is a continuing process over a period of time, the challenge of locating necessary tonnage and grade is thus a continuous one. The production schedule needs to adhere to daily, weekly, monthly and yearly basis. Data obtained during the earlier exploration stages give a broad picture of the deposit, which is enough to give information on localized basis to enable taking up mine planning with production schedules on a reliable basis. Mine exploration aims to help fulfil this data gap.

The twin objectives of mine exploration are: (i) to verify and correct the earlier exploration data, and (ii) control of the mining procedure in grade control, and mine economics.

9.2 Verification and Correction of Earlier Exploration Data

At the stage of mine exploration, more precise data are required which will enable the mining of the ore according to specific grade and tonnage needs. Here, it is important to verify and correct the earlier exploration data. The data need of underground and opencast mines tend to be different nature and therefore each case is dealt with separately.

G.S. Roonwal, *Mineral Exploration: Practical Application*, Springer Geology, DOI 10.1007/978-981-10-5604-8_9

9.2.1 Underground Mines

At the development stage, the data available in the underground mines would be of the following types:

(1) Surface geological plans and sections showing the geology, mineralization, location of mine and exploratory openings, boreholes.
(2) Subsurface composite plane of various levels.
(3) Composite plans showing level details, stope cross-cut, winze, raise and drive positions, borehole intersection points, and
(4) Level-wise or vertical projections of the underground developments.

Where the ore distribution is spotty and in the form of stringers, the data available above may require continuous verification and changes in daily control of ore grade and tonnage. Even in the case of massive ores, modifications are required. Continuous verifications are necessary at the mining stage.

The factors which generally contribute to grade dilution and tonnage difference in underground mines are: (a) imprecise definition of footwall and hanging-wall contacts of the orebody, (b) presence of non-mineralised portions within blocks of ore and (c) the necessity of leaving roof support pillars, etc., within the stopes. These aspects are carried out during geological mapping, sampling, drilling, cross-cutting, raising, winzing and driving and reinterpretation of the consolidated data for small blocks of ore.

Geological mapping: The various underground openings available for detailed examination and mapping are drives, cross-cuts, raises, winzes and stope faces. The details to be recorded are footwall and hanging-wall contacts, strike and dip of all planar features inside and outside the orebody, small-scale structures like folds, faults, shears, contacts between various ore types, barren zones.

The mapping may be done on a 1:200 or 1:100 scale base plan prepared for small sections of the level. The procedure for underground mapping has been discussed earlier and may be followed.

Sampling: Sampling is done to precisely define the assay boundaries between the hanging-wall and footwall contacts and to know the grade of the ore. The boundaries between the mineralized, lean and non-mineralized zones are preciously defined by sampling. A correct grade estimation is also one of the objectives of the sampling. Sampling points and interval of sampling have to be judiciously selected based on thorough inspections and checks. More precise spacing can be done by the various statistical methods discussed earlier in connection with the intensity of exploration. In practice, the sampling may be adopted by cutting channels of 10 cm width and 1–2 cm depth, across the strike of the orebody. Samples are subdivided into 50 cm lengths to enable the data to be treated statistically. In many base metal mines, systematic chip sampling on 15–25 cm grid is being practiced with satisfactory results.

Drilling: The main purpose of drilling at this stage may centre around the guidance of headings, drives, stope definition, stope development, and may be coring or noncoring. Specific grade and tonnage control drilling will have to be undertaken in certain cases. If this is undertaken, spacing of drill holes can be done either based on the experience in such deposits or statistical methods described earlier.

When a procedure based on experience is adopted the exploration and mining data have to be compared and correlated. When the data obtained from a particular grid of density of exploration show close agreement with the mining data in tonnage and grade, then that grid density can be adopted for other blocks. This assumes, of course, the existence of a fair uniformity in the mineral value distribution within the deposit.

Cross Cutting, Raising, Winzing and Driving: In a working mine, such openings will be readily available for observations. However, when exploratory data are generally insufficient or there is some necessity for specific verification or collection of bulk samples, these openings can be made for such exploratory purposes.

Re-examination of data: At time of mining, located data are large for each block/level/stope, so separate grade-wise, type-wise, reserves need to be computed. This is shown in large-scale plans so that daily/weekly/monthly/yearly production schedules can be drawn up. The information available at this stage are:

(a) Borehole intersections of the orebody, both from the surface and underground, (b) Sampling details, (c) Type-wise ore distribution details, (d) Pattern of distribution, ore tonnage and grade in stopes in each level and (e) Details about exploration done in adjacent areas.

Data are processed and stope plans, level plans, assay plans are accomplished and fresh grade-wise reserve assessment is done. Since, production comes mainly from the stopes and partly from the development faces, for purposes grade control, it is necessary that a certain quantum of ore comes from the stope and the rest from developmental headings is collected. The ratio is decided at this stage. Geologist, who conducts the mine level exploration needs to plan strategy to prove it in advance so that the necessary numbers of proved blocks are ready for mining. The geologist has to monitor the rate of depletion from all sections. This is systematically conducted to see correlation between exploration data and the results of mining. At each stage of mine exploration, new data need to be studied and composed with that of earlier exploration to compare the reconcile. This will guide up the formulation of local, mine/section-wise norms for exploration.

9.2.2 Open Pit Mines

Verification and updating is needed in opencast mines as well, its need may arise during development or during mining. The problems and verification procedure of opencast mines are broadly comparable with those of underground mines. At the development and mining stage, the available data in the case of opencast mines would consist of: (a) Surface geological plan and sections showing all details such

Fig. 9.1 Arial view of pit Rampura-Agucha lead zinc mine, India

as borehole, pit, trench position, mine openings, bench locations, etc., on scale of 1:1000, 1:2000 or 1:4000, and (b) Slice plans on larger scales like 1:5000.

Additional exploration shall supplement this data to enable day-to-day grade and tonnage requirements and control. In opencast mines, the grade and tonnage variations occur because of the following factors: (a) Development of waste bands, poor grade ores, etc., between sampling points, boreholes, pits, (b) Unexpected lensing out of high grade ore portions, (c) Imprecise definitions of footwall/hanging-wall contacts with the orebody, and (d) Imprecise definition of overburden, sideburden and interburden.

Verification and re-evaluation at this stage is done by geological mapping, bench sampling, pitting and trenching, drilling by blast hole drills, and reinterpretation of the consolidated data for small ore blocks (Fig. 9.1).

Geological Mapping

The methods described earlier are valid here also. However, the scale would be very large (between 1:500 and 1:200) and the plans will pertain to sections, benches, slices and the details would be different. Some of the details which would require to be mapped are listed below:

(a) Footwall/hanging-wall contacts (observation at every 2–3 m)
(b) Type-wise ore occurrence,
(c) Small-scale features like local folds, faults, shears, etc.
(d) Planar features, like strike, dip foliation joints and
(e) Ore control factors like lithology, colour and mineralogy

During mapping, the data from nearby boreholes, pits, trenches and other exploratory openings will be studied to correlate data with the exposed ore block. This way it is sometimes possible to establish correlation between ore types, ore colours, ore mineralogy and related features with ore grades. Such correlation helps in grade control.

Bench sampling: The usual sampling techniques described earlier are useful here also. However, the objectives may be different. Instead of collecting general samples here, type-wise, colour-wise and mineralogy-wise samples may be preferred to identify and establish the correlation between these physical features and grades. As in mapping, the available exploratory data are studied continuously to establish possible correlation between what is observed in the exploratory openings and exposed ore.

Trenching and pitting: During mine exploration, trenching and pitting would be undertaken on a highly selective basis and for specific purposes like bulk sampling, recovery testing, as described earlier.

Drilling and blast hole drills: During development and exploitation, the ore slices and blocks are drilled at fairly close intervals for blasting purposes. The collection and study of the cuttings generated by this is becoming a useful grade control tool. It is seen that 17 m deep holes of 32 cm diameter generate 3–4 tonnes of cuttings.

The cuttings so generated are allowed to accumulate in a heap around the collar of the drill holes. A trench is dug through the cutting radially from the hole. Some 10% of the cuttings are sliced and removed from one of the trench faces. A sample is now collected just beneath the sliced face by a square nose shovel.

The physical characteristics of the ore cuttings can be studied in the trench cutting. The cuttings will show the strata in reverse order as they are encountered in the drill hole. Each separate layer of cuttings can be measured, studied and correlated with the data available from nearby drill core or ore exposed in benches. By a combination of this study and chemical analysis, the average grade of the ore block can be established just prior to its being blasted. Blocks of blasted ore having specific grades can thus be segregated.

9.2.3 Review of Interpretation of Data

Methods of interpretation have already been discussed. At this stage, however, the data are voluminous and the desired precision of estimate of grade and tonnage is higher. The data available are: (a) Large-scale geological plan of the deposit and mine as a whole, with cross sections, (b) Slice and bench plans, (c) Borehole/pit/trench data and (d) Blast hole data.

From this data, grade-wise and block-wise reserves are marked out for week-wise production planning. Similar to underground mines here, exploration needs of adjacent blocks wise are kept constantly under review, and several blocks of ore are kept proved a little in advance of production.

9.3 Control of Mining Operation

Two important factors to be considered are: (a) grade control (quality control) and (b) mine economics (or out of grade) in the control of mining operations.

9.3.1 *Grade Control*

In all of the mining operations, the mined ore has to be of a specific usable grade. In order to achieve this, grade control operations are necessary to (i) to maintain constant ore quality and (ii) to maintain constant ore size. Before going into the grade control, certain operational mining principles like cut-off grade have to be understood.

Cut-off grade: Cut-off grade is rather broadly defined as the minimum grade of ore which meets the direct operational costs. Cut-off grade is a highly variable quantity being dependent on too many factors, such as market conditions, transportation facilities and ultimately the cost of operations.

Two types of cut-offs have to be recognized. These are planned cut-off and operational cut-off.

Planning cut-off: During exploration, ores have to be defined without taking into account the conditions of mining. In the projection of geological reserves and mineable reserves (during a feasibility study), this cut-off is used. This aim is to predict the total tonnage available in a deposit.

Operational cut-off: When production starts, sectional reserves have to be defined. Such ores may be confined to a particular bench, section, stope, part of a stope, etc. This defines the available ore, concept is based on the principle of the total profit of the mine.

Another concept which is basic to an understanding of grade control is the raw ore grade or the run-of-mine grade. The raw ore is that ore which has just been mined. This contains ores from the highest grades to the cut-off grade and also some very low grades which cannot be practically isolated; quite often, wall rock dilution could also be encountered. This contains ores of highly varying sizes, from large boulders to very small pieces, which may not be more than a few mm in size. Thus, the run-of-the mine ore is a very heterogeneous mixture. In most mines, this does not go untreated to the point of utilization. The final product which comes out of the mine after some treatment is the marketable ore or concentrate which needs to conform to certain size and grade specification.

A block of ore about to be mined normally consists of a variety of ore materials which can be recognized in the following categories: (i) Ores above cut-off grade, (ii) Ores below cut-off grade and (iii) Waste rock.

The mining method may be mechanical or manual or a combination of both. In the manual and semi-mechanized mines, it is not very difficult to mine selectively only those ores which are above the cut-off grade. But, here some low grades and wastes are mined which to dilute ore. In case of mechanized mining, a higher amount of dilution is always anticipated because the machine cannot be as selective in mining as manual agencies. In some of veins mined underground, the width of ore may be less than the minimum stoping width a higher dilution is expected. All factors need to be evaluated prior to grade control have to be collected is attempted. Data are computed and evaluated on the basis of exploration results.

A uniform grade and size of ore is achieved by: (i) By selective mining, or by adopting grade controlled mining sequence for obtaining the required grades, (ii) By blending of mined and sorted ore of different grades at the surface.

Selective mining is possible only in some cases. In many cases, blending may be a viable alternative. For blending, various procedures are available. Those which can be planned prior to mining are discussed here as they alone can be controlled by mine exploration. Some of these alternatives are given below:

(1) Reserving high-grade blocks for blending only. In this, a proven ore block of high grade in a bench or section is exclusively kept to draw ores only for blending with the lower grades of ores. The rate of production in this section will be such that it will last till the last of the low-grade portions are mined out. From a practical stand point, this is difficult.
(2) Having a large number of blocks from which any combination of choices can be made. This needs more practice, one precaution is necessary being the adjustment of the rate of production from a few selected stops or faces.

Such operations call for a very detailed knowledge of the deposit which can be obtained only by intensive exploration. Plans made at this stage have exhaustive details such as:

(1) Distribution of ore types: Physical, mineralogical, colour wise, hardness wise, specific gravity wise, etc.
(2) Distribution of poorly mineralized areas such as: Poor grade ore, ore which may crumble to minable types, etc.,
(3) Distribution of mineralized areas, with lithological details and the quantum which might get mixed up with ore during mining,
(4) Presence of surface/subsurface water, and the possibility of its accumulation in and around the mining areas.

There may be other details which influence mining which require separate attention.

It is often possible to recognize and correlate the grade of ore by one or more of the physical characteristics of the ore type. The presence or absence of a mineral or a group of minerals sometimes controls the ore grade. Colour or weight of a type of ore may be correlated with a specific quality. Such observations should be carefully made so that mining and blending could be planned in advance. A typical blending scheme being worked out in an iron ore mine is discussed below. The planning of such an operation is done on the basis of mine exploration data. The ores in this case have 55–69% Fe and 1.5–10.0% Al_2O_3 with a heterogeneous mixture of hard, soft and friable ores. The blending schedule was worked out on the basis of the following data:

(i) data from plans and sections,
(ii) the quality of ore as determined from blast hole samples,
(iii) the quality of the blasted material as determined from the samples,
(iv) recoveries of lumps and fines from different sections of the mines and
(v) inspection of mine site for observing lithological variations

Blending was done at different stages:

(i) sequencing of trucks from different mine faces (of different ore quality) to the primary crusher,
(ii) travelling of tripper conveyors on the primary surge piles and secondary surge pile at the ore processing plant, and
(iii) using three alternative feeders while loading waggons.

9.3.2 Waste Generation

The economics of any mine operation is dependent on many factors. Of these, the generation and handing of waste material incidental to mining is very important. The cost of production of the ore is closely influenced by the presence of these wastes. A precise estimate of the exact quantum of waste at each stage of mining is an important objective of mine exploration. There are following types of mine wastes: (i) Overburden, (ii) Sideburden, (iii) Interburden, (iv) Undergrade ore and (v) Undersize ore.

(i) Overburden: Overburden is defined as any non-ore material covering the orebody, which has to be removed to expose the ore. The constitution of the overburden maybe anything from loose soil to hard rock. It may also be laterites, low-grade ore materials, etc. In many cases, the overburden has to be stripped out completely before any mining can commence.
(ii) Sideburden: Sideburden is essentially overburden but is recognized separately depending upon the attitude of the orebody. If the orebody is steeply inclined, then sideburden may occur on the hanging-wall or footwall side or on both sides and is required to be removed before extracting the ore.
(iii) Interburden: Interburden is any waste rock which occurs within two mineable ore bodies or two or more mineable parts of an orebody.
(iv) Undergrade ore: Ore which has to be mined but cannot be sold because of its grade being below the cut-off grade is considered as undergrade or sub-marginal grade and is considered a mine waste unless otherwise used for beneficiating the same.
(v) Undersize ore: Like undergrade ore there may be under size ore which cannot be utilized due to its small size. This also forms a mine waste unless it is used for beneficiation and/or agglomeration.

During the estimation of mineable reserves, all factors have to be taken into consideration. The market value of every tonne of ore should absorb along with other costs the cost of removing any or all these wastes. The ore to waste ratio is the most important aspect of mine economics in the case of opencast mines. In the case of underground mines, a major factor is the inevitable overbreakage and underbreakage along the footwall and hanging walls which influence ore dilution, which in turn influence mine economics and cut-off grade. When the cost of handling waste/undersize/undergrade material is not effectively absorbed by the market price of the ore, mining becomes uneconomic.

In open pit mines, this economic limit is forecast on the basis of exploration data by projecting the ultimate pit limit on cross-section plans or slice plans. Consideration of the ultimate pit slope is also very important in these projections. The ultimate pit slope is determined by considerations of safety as well as the stability of the rocks involved.

In underground mines, the economic limit is guided among other things by the minimum stoping width. This is applicable in the case of ore bodies which are thin. The minimum stoping width defines the narrow stope width in which miners can expeditiously. If the orebody happens to be thin than this width, mining of the necessary extra width on both the sides will produce waste and poor grade ore which will dilute the ore grade and hence adversely affect the mine's economics. In mine exploration, such situations should be anticipated early by advance face exploration.

9.3.3 Geotechnical Investigation in Open pit Mining

Geotechnical investigations are required for large-scale open-cast mining. Although such investigations are essentially engineering studies, geological factors form important basic data required. Three types of geotechnical studies require geological data. They are stereographic projection (plans, features, joints, minor fold axes, cleavage), physical scale models and finite element analysis.

1. Stereographic projection: The planar feature studies may be of joints, cleavages, foliation, minor fold axes, etc. The edge of the friction of each joint is studied to select the most critical wedges. These data combined with the geological map provide information about the potential areas of failure.
2. Physical models: These are made to select homogeneous units which are manageable as a unit in the context of slope stability. For this, the areas are studied and classified into homogeneous structural units.
3. Finite element analysis: For this, the essential data are stress strain coefficients of each rock unit. Anisotropy of rocks can be studied this way. Tensile strength and shear strength values of each rock unit are also studied.

Further Readings

Arogyaswami RNP (1988) Course in mining geology, 3rd ed Oxford IBH, p 695

Burn RG (1984) Exploration Risk. CIM Bu 77(870):55–61

Laurence D (2011) A guide to leading practice sustainable development in mining. Australian Government, Department of Resource, Energy and Tourism. http://lwww.rel.go.\au, p 198

Appendix A

Conversion factors

1 m	=	3.281
1 cm	=	0.39 in.
1 μm	=	0.039
1 foot	=	0.305 m
1 km	=	100 m = 0.621 miles
1 mile	=	1.609 km
1 g	=	0.035 ounce (dry)
1 ounce (dry)	=	28.35 g
1 kg	=	1000 g = 2.205 pounds
1 pound	=	0.454 kg
1 t	=	0.984 kg long tonne (UK) = 1.102 short tonne
1 long tonne (UK)	=	1.016 t = 1.120 short tonnes (US)
1 short tonne (US)	=	0.507 t = 0.893 long tonne (UK)
1 l	=	0.220 gallon UK = 0.264 gallon (US)
1 gallon UK	=	4.546 l = 1.201 gallon US
1 gallon US	=	3.785 l = 0.833 gallon UK
1 barrel (oil)	=	0.132 t = 0.134 long tonnes (UK)
1 barrel (oil)	=	35 gallon (UK) = 42 gallon (US)
High numerals		
1,000,000	=	Million
1,000,000,000	=	Billion (1000 million in Europe)
1,000,000,000,000	=	Trillion (1000 billion in Europe)
Area		
10,000 m^2	=	1 ha (2.471 acres)
1 acre	=	0.405 ha
1 $mile^2$	=	2.599 km^2

G.S. Roonwal, *Mineral Exploration: Practical Application*,
Springer Geology, DOI 10.1007/978-981-10-5604-8

Appendix B

Properties of import economic minerals

Mineral	Chemical composition	Specific gravity	Hardness (Moh's scale)	Colour	Streak	Lustre
1	2	3	4	5	6	7
I. Native elements						
Antimony	Sb	6.7	3–3.5	Tin-white	Tin-white	Metallic
Arsemoc	As	5.7	3.5	Tin-white	Tin-white	Metallic
Copper	Cu	6.6–8.9	2.5–3	Copper-red	Copper-red	Metallic
Diamond	C	3.5–3.53	10	White or colourless	–	Adamantine
Gold	Au	15.6–19.3	2.5–3	Gold-yellow	Gold-yellow	Metallic
Graphite	C	2.09–2.23	1–2	Iron-black		Metallic or earthy
Lead	Pb	11.4	1.5	Lead-grey		Metallic
Mercury	Hg	13.6	–	Tin-white		Metallic
Platinum	Pt	14–19.9	4–4.5	Steel-grey		Metallic
Silver	Ag	10.1–19.19	2.5–3	Silver-white		Metallic
Sulphur	S	2.05–2.09	1.5–2.5	Yellow	White	Resinous
II. Sulphides, selenides, tellurides, etc.						
Realgar	AsS (As-70.1%)	3.56	1.5–2	Yellow-orange-red	Orange-red	Resinous
Orpiment	As_2S_3 (As-61%)	3.4–3.5	1.5–2	Lemon-yellow	Lemon-yellow	Pearly
Stibnite	Sb_2S_3 (Sb-71.7%)	4.52–4.62	2	Lead-grey	Lemon-grey	Metallic
Bismuthnite	Bi_2S_3 (Bi-81.2%)	6.4–6.5	2	Lead-grey	Lemon-grey	Metallic
Molybdenite	MoS_2 (MO-60%)	4.7–4.8	1–1.5	Lead grey	Greenish grey	Metallic
Galena	PbS (Pb-86.6%)	7.4–7.6	2.5–2.75	Lead-grey	Lemon-grey	Metallic
Argentite	Ag_2S (Ag-87.1%)	7.2–7.36	2–2.5	Lead-grey	Lemon-grey	Metallic
Chalcocite	Cu_2S (Cu-79.8%)	5.5–5.8	2.5–3	Lead-grey	Lemon-grey	Metallic
Sphalerite	ZnS (Zn.6996)	3.9–4.1	3.5–4	Reddish-brown black to white	Brownish	Resinous to adamantine
Cinnabar	HgS (Hg-86.2%)	8–8.2	2–2.5	Brownish-red	Scarlet	Adamantine to metallic
Greenockite	CdS (Cd-77.7%)	4.9–5	3–3.5	Orange-yellow	Orange-yellow	Adamantine to resinous
Millerite	NiS (Ni-64.7%)	5.3–5.65	3–3.5	Brass-yellow	Greenish	
Niccolite	NiAs (Ni-43.9%)	7.33–7.67	5–5.5	Copper-red	Brownish-black	
Pyrrhotite	Fe_3S_6–$Fe_{16}S_{17}$	4.58–4.64	3.5–4.5	Copper-red	Greyish-black	
Covellite	CuS (Cu-66.4%)	4.6	1.5–2	Indigo-blue	Lead-grey	
Bornite	Cu_3FeS_4 (Cu-66.4%)	4.9–5.4	3	Brown or copper-red	Greyish-black	Metallic
Chalcopyrite	$CuFeS_4$ (Cu-34.5%; Fe-30.5%)	4.1–4.3	3.5–4	Brass-yellow	Greenish black	Metallic

(continued)

G.S. Roonwal, *Mineral Exploration: Practical Application*,
Springer Geology, DOI 10.1007/978-981-10-5604-8

(continued)

Mineral	Chemical composition	Specific gravity	Hardness (Moh's scale)	Colour	Streak	Lustre
1	2	3	4	5	6	7
Pyrite	FeS_2 (S-53.4%; Iron-46.6%)	4.95–4.97	6–6.5	Silver-white to red	Greenish or brownish-black	Metallic
Cobaltite	CuAsS (As-45.2%)	6.63	5.5	Silver-white to red	Greyish-black	Metallic
Arsenopyrite	FeAsS (As-46%; Iron-34%)	5.9–6.2	5.9–6.2	Steel-grey	Dark greyish-black	Metallic
III. Sulpho-salts						
Bournonite	2 $PbsCu_{25}$, Sb_2S (Pb-42.5%, Cu-13%, Sb-24.7%)	5.7–5.92	2.5–3	Steel-grey	Steel-grey	Metallic
Pyrargyrite	3 Ag_2S Sb_2S_3 (Ag-59.9% Sb-22.3%)	5.77–5.86	2.5	Black to greyish-black	Purplish-red	Metallic-adamantine
Tetrahedrite	$(Cu, Fe)_{12}$ Sb_4 S_{13} (Cu-52.1% Sb-24.8%)	4.4–5.1	3–4.5	Steel-grey and Iron-black	Grey to iron black	Metallic
Stannite	Cu_2S Fe_5SnS Iron-13.1%) (Sn-27.5%, Cu-29.5%)	4.3–4.5	3.5	Iron-black	Black	Metallic
IV. Haloids						
Halite	NaCL (Na-39.4%; Cl-60.6%	2.1–2.6	2.5	Colourless white, reddish yellowish	Colourless	Vitreous
Cerargyrite	AgCl (Ag-75.3%)	5.8	2–3	Pale-grey, greyish-green, bluish, colourless	Shining	Vitreous
Fluorite	CaF_2	3.01–3.25	4	White, yellow, green, rose, brown, blue	White	Vitreous
Cryolite	Na_3AIF5	2.95–3	2.5	White, colourless reddish, brownish	–	Vitreous to greasy
Corundum	Al_2O_3	3.95–4.10	9	Blue, red, grey yellow brown	Colourless	Vitreous to adamantine
Haematite	Fe_2O_3 (Iron-70%)	4.9–5.3	5.5–5.5	Iron-black	Cherry-red	Metallic
Ilmenite	FeO_2 TiO_2 (Ti-31.6% iron 36.8%)	4.5–5	5.6	Iron-black	Brownish-red	Submetallic
Magnetite	FeO_2, Fe_2O_3 (Iron-72.4%)	5.17–5.18	5.5–6.5	Iron-black	Black	Metallic
Chromite	Feo-Cr_2O_3 (Cr-46/5%)	4.1–4.9	5.5	Iron-black to brownish-black	Brown	Submetallic to metallic
Chrysoberyl	BeO, Al_2O_3 (Alumina-80.2%)	3.5–3.84	8.5	Shades of green	Colourless	Submetallic
Braunite	$3Mn_2MnO_3$ $MnSio_3$	4.75–4.82	6–6.5	Steel-grey to brownish-black	Brownish-black	Submetallic
Cassiterite	SnO_2 (Sn-78.6%)	6.8–7.1	6–7	Brown-black	White, grey, brownish	Adamantine
Rutile	TiO_2 (Ti-60%)	4.18–5.2	6–6.5	Reddish-brown, red, yellow	Pale brown	Metallic to adamantine
Pyrolusite	MnO_2 (Mn-63.2%)	4.73–4.86	2–2.5	Steel-grey to iron-black yellow, violet, bluish	Black to bluish black	Metallic
Diaspore	$Al_2O_3H_2O$ (Alumina-85%)	3.5	6–7	Colourless, white brown, yellow	–	Vitreous to pearly
Goethite		4.28	5–5.5		Brownish-yellow	

(continued)

(continued)

Mineral	Chemical composition	Specific gravity	Hardness (Moh's scale)	Colour	Streak	Lustre
1	2	3	4	5	6	7
	$Fe1O_3$ H_2O (Fe-62.9%)			Yellow, red, blackish brown		Imperfect adamantine
Manganite	Mn_2O_3, H_2O (Mn-62.4%)	4.2–4.4	4	Iron-black	Reddish-brown	Submetallic
Limonite	$2Fe_2O_3$ $3H_2O$ (Fe-59.8%)	3.6–4	5–5.5	Shades of brown	Yellowish-brown	Submetallic
Bauxite	Al_2O_3 $2H_2O$	2.55	-	White, grey, yellow, brownish, reddish-brown	–	Earthy
Gibbsite	Al_2O_3 $3H_2O$ (Al-65.4%)	2.3–2.4	2.5–3.5	White, grey, greenish	–	Pearly to vitreous
Psilomelane	$4MnO_2$ H_2O (Ba_2K_2)	3.3–4.7	5–7	Iron-black	Brownish-black	Submetallic
VI. Oxygen-salts **(I) Carbonates**						
Calcite	$CaCO_3$ (Lime-56%)	2.71	3	Colourless, white, red, green, blue, yellow	Greyish	Vitreous to earthy
Dolomite	$CaCO_3MgCO_3$ (Lime-30.4%) (MgO-21.7%)	2.8–2.9	3.5–4	White, reddish rose-red, green, brown, black	–	Vitreous to earthy
Magnesite	$MgCO_3$ (Magnesia-47.6%)	3–3.12	3.5–4.5	White, yellow, brown	White	Vitreous
Siderite	$FeCO_3$ (Fe-48.2%)	3.83–3.88	3.5–4	Grey, yellow, green, brown, red	White	Vitreous to pearly
Rhodochrosite	MCO_3 (Mn-47.8%)	3.5–3.5	3.5–4.5	Rose-red, yellow, brown	White	Vitreous to pearly
Smithsonite	$ZnCO_3$ (Zn-52.1%)	4.3–4.45	5.5	Greyish, brown, green, blue	White	Vitreous to pearly
Aragonite	$CaCO_3$ (Lime-56%)	2.93–2.95	3.5–4	White, grey, yellow, green, violet	Colourless	Vitreous
Witherite	$BaCO_3$ (Baryta-77.7%)	4.27–4.35	3–3.75	Greyish, yellowish	White	Vitreous
Strontianite	$SrCO_3$ (Strontia-70.1%)	3.68–3.71	3.5–4	Green, grey, yellow, brown	White	Vitreous
Cerussite	$PbCO_3$ (Lead oxide-83.5%)	6.5–5.6	3–3.5	Grey, blue, green	Colourless	Adamantine to pearly
Malachite	$CuCO_3$ $(Cu(OH)_2$ (Cupric oxide-71.9%)	3.9–4.0	3.5–4	Bright-green	Pale-green	Adamantine to Vitreous
Azurite	2 $CUCO_3$, Cu $(OH)_2$ (Cupric oxide-69.2%)	3.77–3.89	3.5–4	Azure-blue	Blue	Adamantine to Vitreous
(2) Silicates						
Orthoclase	$K_2OAl_2O_2$ $6SiO_2$ (Potash-16.9%)	2.51–2.58	6	Colourless, white, yellow, red	Colourless	Vitreous
Microcline	K_2O Al_2O_3 $6SiO_2$ (Potash-16.9%)	2.60–2.62	6–6.5	White, yellow, red, green	Colourless	Vitreous
Albite			6–6.5		Colourless	Vitreous

(continued)

(continued)

Mineral	Chemical composition	Specific gravity	Hardness (Moh's scale)	Colour	Streak	Lustre
1	2	3	4	5	6	7
	Na_2O Al_2O_3 $6SiO_2$ (Soda 11.8%). (Alumina-19.5%)	2.60–2.62		White, blue, red, green		
Oligoclase	NaSl SiO_2O_3	2.6–2.75	6–6.5	White, green, yellow	Colourless	Vitreous
Andesine	Na $AlSi_3O_3$			White or grey		
Labradorite	$CaAlSi_3O_3$	2.67	6	Grey, dark, ashy brown		
Bytownite	$NaAlSi_3O_5$	2.72		Grey, dark grey, bluish		
Anorthite	CaO Al_2O_3 $2SiO_2$	2.74–2.76	6–6.5	Grey, red	Colourless	Vitreous
Leucite	KAl $(SiO_3)_2$	2.45–2.5	5.5–6	Grey	Colourless	Vitreous
Enstatite	$MgSiO_3$	3.1–3.3	5.5	Grey, yellow, green, brown, black brownish grey	Colourless	Vitreous
Hypersthene	(FeMg) SiO_3	3.4–3.5	5–6	Black, brownish, green		
Diopside	Ca Mg $(SiO_3)_3$	3.2–3.38		Black, white, green	Greyish	Pearly
Augite	Ca, Mg SiO_2 O_5	3.2–3.6	6.6	Black, brown	White	Pearly
Spodumene	Li_2O, Al_2O_3, 4 SiO_2	3.13–3.2	6.5–7	Green, white, yellow, purple	White	Pearly
Rhodonite	MnO, SiO_2	3.4–3.68	5.5–6.5	Red, brown, pink, green	White	Vitreous
Wollastonite	CaO, SiO_2	2.8–2.9	4.5–5	Grey, yellow, red, brown	White	Vitreous
Anthophyllite	(Mg, Fe) SiO_2	2.85–3.2	5.5–6	Brown, grey, greenish, grey, green	Colourless	Vitreous
Hornblende	Ca (Mg Fe)$_3$ Si_4O_{12} Ca Mg_2 (Al Fe)$_2$ Si_3O_{12}	2.9–3.4	5–6			Vitreous
Grossularite	3 CaO Al_2O_3 3 SiO_2	3.53	7	Colourless white, green, yellow	White	
Almandine	3 FeO Al_2 O_2	4.25	7	Red	White	Vitreous to resinous
Pyrope	3 MgO Al_2O_3 3 SiO_2	3.51	7	Deep red black	White	Vitreous to resinous
Spessartite	3 MnO Al_2O_3	4.18	7	Red, brownish-red	White	Vitreous to resinous
Andradite	3 CaO Fe_2O_3 3 SiO_2	3.75	7	Yellow, green, brown, black	White	Vitreous to resinous
Uvarovite	3 CaO Cr_2O_3 3 SiO_2	3.4–3.5	7.5	Emerald-green	White	Vitreous to resinous
Zircon	Zn SiO_4 (Zirconia-67.2%)	4.68–4.7	7.5	Colourless, yellow, brown, red	Colourless	Adamantine
Topaz	$(AIF)_2$ SiO_3	3.4–3.6	8	Yellow, ray, blue, red, green	Colourless	Vitreous
Andalusite	Al_2SiO_3	3.16–3.2	7.5	White, red, violet, brown, green	Colourless	Vitreous
Kyanite	Al_2SiO_3	3.56–3.67	5–7.25	Blue, white, grey, green, black	Colourless	Vitreous
Sillimanite	Al_2SiO_3	3.23–3.24	6–7	Brown, grey, greenish	Colourless	Vitreous
Epidote	H_2O 4Cao	3.2–3.5	6–7		Colourless	Vitreous

(continued)

(continued)

Mineral	Chemical composition	Specific gravity	Hardness (Moh's scale)	Colour	Streak	Lustre
1	2	3	4	5	6	7
	$3(AlFe)_2\ O_3$			Yellowish-green, brownish, black, greyish		
Tourmaline	$HgAl_2\ (B.OH)_2\ SiO_4\ O_{19}$	2.98–3.2	7–75	Black, brownish, blue, green, red	Colourless	Vitreous
Staurolite	$2\ (Al_2\ SiO_5)\ Fe\ (OH)_2$	3.7	7–7.5	Reddish-brown, yellowish-brown, brownish-black	Colourless	Subvitreous to Resinous
Muscovite	$(HK)\ Al\ SiO_2$	2–2.25	2.76–3	Colourless, grey, brown, green, violet, rose-red	White	Vitreous to pearly
Lepidolite	$(OH\ F)_2Kli\ Al_2Si_3O_{10}$	2.8–3.3	2.5–4	Rose-red, lilac yellowish, white	Colourless	Pearly
Biotite	$K_2K\ (Mg\ Fe)_3\ Al\ (SiO_4)_3$	2.7–3.1	2.5–3	Green to black brown	Colourless	Pearly
Phlogopite	$H_2K\ Mg_3\ Al\ (SiO_4)_3$	2.78–2.85	2.5–3	Brown, yellow, green, white colourless	Colourless	Pearly
Serpentine	$H_4\ Mg_3\ SiO_2$	2–2.65	5–5	Green, brownish, red	Colourless	Pearly
Talc	$H_2\ Mg_2\ (SiO_3)_4$	2.7–2.65	1–1.5	Apple-green, white, brownish, reddish	Colourless	Pearly
ChrysocollaColumbite	$Cu\ SiO_3\ 2H_2O$	2–2.24	2–4	Greenish, blue	Colourless	Vitreous
Tantalite	$O\ (Fe\ Mn)\ (Nb\ Ta)$	5.3–7.8	6	Grey, brownish-black	Dark red to black	Submetallic
Samarskite	$R_3R_2\ (Nb\ Ta)_6\ O_{21}$	5.6–5.8		Velvet black	Dark reddish brown	Vitreous to resinous
Monazite	$(CeLaDi)\ PO_4$	5.27	5–5.5	Red, brown, red, yellowish	White	Resinous
Apatite	$3\ Ca_3\ P_2\ O_8\ Ca\ (Cl\ F)_2$	3.17–2.23	4.5–5	Green, violet, blue, white, brown, red	White	Vitreous, subresinous
Vanadinite	$(PbCl)\ Pb_4\ (VO_4)_3$	6.66–7.1	2.75–3	Ruby-red, brown, yellow	White or yellowish	Resinous
Lazulite	$2\ Al\ PO_4\ Fe\ Mg\ (O_4)_3$	2.6–2.83	5–6	Azure-blue	White	Vitreous
Turquois	$CuO_3\ Al_3O_3\ 2P_2O_4\ HO$	2.6–2.83	5–6	Blue, apple-green, or greenish	White	Waxy
Barytes	$BaSO_4$	3–4.6	2.5–3.5	White, yellow, blue, red, brown	White	Vitreous to resinous
Celestite	$SrSO_4$	3.95–3.97	3–3.5	White, blue, red	White	Vitreous
Anglesite	$ObSO_4$	6.3	3	White, yellow, green, blue	Colourless	Adamantine to vitreous
Gypsum	$CaSO_4\ 2H_2O$ (Lime-32.5%)	2.3	1.5–2	Grey, white	White	Pearly
Wolframite	$(FeMn)\ WO_4$	7–7.5	5–5.5	Greyish, brownish	Black	Submetallic to Adamantine
Scheelite	$CaWO_4$	6	4.5–5	White, yellow, brown, green, red	White	Submetallic to Adamantine
Wulfenite	$Pb\ MO\ O_4$	6.7–7	3	Orange-yellow, olive-green, grey, brown, red	White	Submetallic to Adamantine

Appendix C

Assessment of potential orefields
Geological factors principal metals, grade (i.e. concentration) of ore
Possible by products—(e.g. silver as trace metal enhances value of PbZn deposits)
Size and structure of are body estimated reserves
Mining and metallurgical considerations mining procedure necessary, e.g. opencast/underground
Extraction processes required (e.g. ease of separation of ore minerals, suitability for cheap treatment of purified ore)
Suitability of end product (e.g. low levels of impurities such as Pin iron are desirable for use in steel making)
Other considerations price of metal on world market expected future demand
Ease of access cost of development legislation relating to mining
Likelihood of future changes in social, political or economic climate liable to affect development

G.S. Roonwal, *Mineral Exploration: Practical Application*,
Springer Geology, DOI 10.1007/978-981-10-5604-8

Appendix D

Metalliferous mineral deposits: a geological grouping

(a) Deposits related to igneous processes orthomagmatic: segregated during consolidation Pneumatolytic: associated with residual fluids exhalative and fumarolic: deposits of volcanic centres some hydrothermal deposits (see (d))
(b) Deposits related to sedimentary processes placers and related deposits: segregated by physical Processes metalliferous chemical sediments including deep-sea Deposits metalliferous residual sediments and weathering products some hydrothermal deposits (see (d))
(c) Deposits related to metamorphic processes pyrometasomatic deposits: formed at igneous contacts other metasomatic deposits of groups (a), (b) and (d) modified by metamorphism
(d) Hydrothermal deposits: formed through the agency of hot waters circulating in the host rocks: vein deposits replacement deposits

G.S. Roonwal, *Mineral Exploration: Practical Application*,
Springer Geology, DOI 10.1007/978-981-10-5604-8

Appendix E

Metals and their sources

	Principal types of deposit	Approximate annual production (1977) (tonnes)
Precious metals gold	Hydrothermal, placer	1200
Silver	Volcanic, hydrothermal	70,000
Platinum	Orthomagmatic	700
Light metals aluminium	Residual sedimentary hydrothermal	14 million
Steel industry metals		
Iron	Sedimentary	850 million
Nickel	Magmatic, residual	500,000
Manganese	Sedimentary	10 million
Chromium	Orthomagmatic	9 million
Cobalt	Sedimentary	20,000
Molybdenum	Hydrothermal, volcanic	70,000
Tungsten	Pneumatolytic	45,000
Non-ferrous metals		
Copper	Volcanic, sedimentary	6 million
Tin	Pneumatolytic, placer	200,000
Zinc	Volcanic, hydrothermal, sedimentary	4 million
Lead	Volcanic, hydrothermal, sedimentary	3.5 million

G.S. Roonwal, *Mineral Exploration: Practical Application*,
Springer Geology, DOI 10.1007/978-981-10-5604-8

Appendix F

Exploration procedures: an idealized sequence (for details of techniques see Chap. 8)
Decision to prospect, based on: regional geology *favours* mineralization of desired type, *evidence* of mining in former times random finds of gossan or 'shows' of sub-economic mineralization
Regional reconnaissance by: airborne geophysical surveys (magnetic, electromagnetic, radiometric, gravimetric) geochemical reconnaissance (stream-sediment or lake-sediment sampling) photogeological survey. Including use of satellite imagery ground geological reconnaissance, e.g. traverse mapping leading to: identification of favourable anomalies. Staking of claims; random finds, staking of claims; or rejection of unfavourable areas, abandonment of project
Investigation of selected target areas: geological mapping, ground geophysical surveys (gravimeter, magnetometer, resistivity and induced potential surveys), detailed geochemical surveys (closely spaced, stream-sediment sampling. sampling of drift, soil), exploratory pitting, trenching, trial boreholes, leading to: discovery of deposits. staking of claim; identification of probable buried ore body; or abandonment of project
Assessment of ore body: detailed topographic and geological survey of site, further boreholes, logging of cares petrographical and chemical study of cores, assaying of are samples, leading to: decision to develop; decision to suspend operations; or decision to relinquish claim

G.S. Roonwal, *Mineral Exploration: Practical Application*,
Springer Geology, DOI 10.1007/978-981-10-5604-8

Appendix G

Ore deposits of igneous affinities

Igneous assemblage and tectonic setting	Dominant ore-forming processes	Forms and characters of deposits	Principal metals
(a) Mid ocean ridge volcanic (constructive plate margins, ophiolites)	(i) Exhalative and hydrothermal in lavas	Stratabound massive sulphides formed on sea floor, or veins and replacement bodies	Cu (Ni)
	(ii) Orthomagmatic in ultrabasic intrusions	Layers or irregular bodies	Cr
(b) Greenstone belts (early Precambrian, mainly basic volcanic)	(i) Exhalative	Massive or disseminated sulphides, stratiform iron formations	Cu Zn Fe Au
	(ii) Orthomagmatic	Chromite layers or sulphide segregations with Ni	Ni Cu Cr
	(iii) Hydrothermal	Quartz veins, carrying gold scavenged from volcanic by hydrothermal fluids	Au
(c) Island arcs and orogenic mountain belts	(i) Exhalative Kuroko type	Stratabound sulphides in volcanic usually near rhyolites	Cu (Zn–Pb)
	(ii) Hydrothermal processes round subvolcanic acid stocks–porphyry coppers	Disseminated low-grade sulphides in brecciated roof of stock	Cu (Mo) CuZnPbAu
	(iii) Hydrothermal processes mainly in subvolcanic zone	Polymetallic sulphides, disseminated and in veins	Ag Sb Mo
	(iv) Pneumatolytic and hydrothermal in an round acid stocks and subvolcanic intrusions (mainly of Phanerozoic age)	Cassiterite-bearing quartz veins and pegmatite, disseminated ores sometimes associated with tourmalinization; for uranium see Sect. 3.4	Sn (W Nb) U
(d) Continental rift valleys	(i) Hydrothermal in plateau basalts	Native copper or sulphides in vesicles or permeable horizons	Cu

(continued)

G.S. Roonwal, *Mineral Exploration: Practical Application*, Springer Geology, DOI 10.1007/978-981-10-5604-8

(continued)

Igneous assemblage and tectonic setting	Dominant ore-forming processes	Forms and characters of deposits	Principal metals
	(ii) Orthomagmatic in and around carbonatities or alkali complexes	Stratiform chromite layers or segregations of sulphides	Cr Pt Ni Cu
	(iii) Metasomatic in and around carbonatites or alkali complexes		Nb, REE
(e) Oceanic islands intraplate oceanic	Little important mineralisation		
(f) Continental cratons	(i) As for (d)(i)		Cu Ni
	(ii) As for (d)(ii)		Cr Pt Ti Fe
	(iii) Orthomagmatic in	Stratiform layers of ilmenite-magnetite	Fe Ti
	anorthosites (mainly of Proterozoic age)		
	(iv) Pneumatolytic around granite stocks for kimberlite see Chap. 5, diamond	Vein deposits carrying cassiterite	SnNb

Appendix H

Ore deposits formed by surface processes

Host rocks and environment	Dominant ore-forming processes	Forms and characters of deposits	Types of deposit and examples	Metals
(a) Rocks in zone of weathering and residual	(i) Concentration in situ by removal of soluble components, especially in tropical zones (ii) Solution in oxidized zone deposition near water table	Irregular layers and patches at surface or below unconformities Zones of secondary enrichment of sulphide ore bodies at surface or below unconformities	Bauxite, Jamaica; silicate nickel, New Caledonia; laterite, tropical Africa	Al Ni Fe and as building material
(b) Detrital sediments	(i) Mechanical segregation of heavy minerals	In disaggregated weathered mantle (eluvial), as concordant layers or lenses in fluviatile and beach deposits	Places: Orange River, Namibia (diamonds); Queensland coast (monazite); California (gold)	Au Sn diamond monazite
	(ii) As (b)(i), but heavy mineral concentration diagenetically reworked	Ancient places represented by pebble conglomerates or quartzites	Witwatersrand (South Africa); Blind River, Ontario	Au U U
	(iii) As above	Stratiform, lensoid	Sullivan, British Columbia, Broken Hill, New South Wales	Pb Zn
	(iv) Deposition from pore fluids during or after diagenesis, sometimes under reducing conditions	Lensoid and irregular bodies	'Red-bed' copper deposits, Colorado plateau; roll-front uranium (see Sect. 3.4)	

(continued)

G.S. Roonwal, *Mineral Exploration: Practical Application*, Springer Geology, DOI 10.1007/978-981-10-5604-8

(continued)

Host rocks and environment	Dominant ore-forming processes	Forms and characters of deposits	Types of deposit and examples	Metals
(c) Argillaceous sediments	Adsorption on clay particles, reduction of sulphate by bacterial action	Usually associated with carbonaceous pelites	Kupferschiefer (Permian) of Germany	Cu (Zn–Pb)
(d) Carbonate rocks and evaporates	Deposition from circulating saline waters	Irregular replacement bodies, veins, breccias fillings, often concentrated in limestone reefs, dolomites or evaporitic horizons	Mississippi Valley type ores	Pb Zn
(e) Mixed host rocks	(i) Facies control near algal reefs	Irregular stratiform	Copper Belt, central African Mount Isa, Queensland	Cu Co
	(ii) Acid volcanicity in marine basin	Lensoid		Pb Zn Cu
(f) Chemical	(i) Precipitation and diagenetic reworking, aided by organic activity	Banded iron formations; deposits with chert and/ or carbonate layers: entirely Precambrian	Lake Superior, Canada; Hamersley Range, W. Australia	Fe (mainly as oxides)
(g) Chemical sediments	(i) Precipitation in nearshore environments, diagenetic reworking	Ironstones, bedded deposits, often oolite		
	(ii) Precipitation in shallow, partly enclosed sea	Concordant lenses interbedded with detrital sediments and limestones	Nikopol type, north of Black Sea	Mn (oxides carbonates)
	(iii) Precipitation in deep ocean basins	Nodular bodies and encrustations on deep-sea floor	Manganese nodules, Pacific Ocean, especially 0-20°N of Equator	Mn (Fe) Ni
	(iv) Exhalative metalliferous sediments; see Section 4.3.4			

Appendix I

Hydrothermal vein deposits

Vein assemblage	Common minerals	Geological setting	Examples	Principal metals
Precious metals	Gold tellurides, native gold, quartz, sulphides	Volcanic provinces, especially early Precambrian greenstone belts and Tertiary circum-Pacific provinces T 600–100 °C	Cripple Creek, Colorado (Tertiary) Kirkland Lake, Ontario (Precambrian)	Au Ag
Polymetallic sulphides	Sulphides, quart, carbonates, tourmaline	In subvolcanic settings and around granitic plutons T 500–100 °C	Western USA, SW England (Hercynian)	Some or all of: Cu Zn Pb Ag Au Fe Mo (W Sn)
Lead-zinc	Sphalerite-galena, quartz, fluorite, barite, calcite	In carbonate hosts not directly related to igneous centres $T < 200$ °C	Mississippi-Valley-type ores	Pb Zn
Cobalt type silver	Sulphides, arsenides, native silver, bismuth quartz, calcite	Rare, near basic igneous intrusions	Cobalt district Ontario (Precambrian)	Ag Co Ni Bi

G.S. Roonwal, *Mineral Exploration: Practical Application*,
Springer Geology, DOI 10.1007/978-981-10-5604-8

Appendix J

Exploration methods: summary

Geophysical methods	Principal applications
Seismic surveys (reflection and refraction)	Elucidation of subsurface structure, especially structure of sedimentary basins. Recognition of key horizons, unconformities, folds, faults in oil and gas fields: exploration of superficial deposits of construction sites
Seismicity records	Monitoring of active volcanic centres, fault zones
Gravity	Elucidation of regional structure in sedimentary, igneous and metamorphic terrains. Useful at reconnaissance stage of exploration for minerals. Hydrocarbons: identification of anomalies related to buried igneous centres, ore deposits, salt domes (oil traps)
Magnetic	Elucidation of regional structure in igneous and metamorphic terrains, useful for reconnaissance for mineral deposits; identification of anomalies related to buried igneous centres, iron formations
Electrical and electromagnetic	Resistivity surveys, principally to locate ore bodies and in borehole logging self potential (SP) and induced potential (IP) surveys to locate ore bodies electromagnetic surveys to locate ore bodies and detect rise of magma in volcanic centres
Radiometric	Prospecting for uranium, thorium; borehole logging investing Cltion of areas with anomalous radioactivity
Remote sensing	Airborne geophysical reconnaissance: regional structure topographical surveys geological reconnaissance surveillance of potential hazards such as volcanic centres monitoring environmental changes
Subsurface sampling boreholes	Elucidation of regional succession and structure; mud logging for general lithology, microfauna (oilfields) logging of core for structural and petrogenetic detail (oilfields, mineral exploration, site investigation)
Augering	Investigation of weathered mantle, superficial deposits

(continued)

G.S. Roonwal, *Mineral Exploration: Practical Application*,
Springer Geology, DOI 10.1007/978-981-10-5604-8

(continued)

Geophysical methods	Principal applications
Geochemical and mineralogical methods reconnaissance surveys	Stream- or lake-sediment samples, water samples, reconnaissance for ore deposits, basis for investigation of geochemistry in relation to plant, animal or human health heavy mineral concentrates
Local surveys	Location of mineral deposits, investigation of possible geochemical hazards
Periodic sampling	Quality control (water and effluents): effects of pollution

Appendix K

Mineral Belts of India

All geological formations, however, do not contain mineral deposits. There is a relationship between formation of major rock units tectonic and ore genesis as explained in Chap. 2. In the Indian context, three major tectonic units which include structural metallogenic zones: (i) shield areas, (ii) mobile belts, and (iii) platform areas. In addition to these three units, there are two other areas of interest for mineral explorations: (i) the lateritic belts and (ii) the Quaternary formations such as alluvium, loess, beach sands. Table 10.1 gives major geological formation of India.

The formations such as the Gondwana, the Deccan Trap, the Cretaceous, the Triassic, the Jurassic, and the whole of the Tertiary have not indicated significant mineralization. Limestone, phosphatic nodules, building stones, China clay, fire-clay, gypsum are present in some of them; the sedimentary and evaporate type of deposits are considered in another section. Coal and oil, which are present in some of the formations, are not considered here.

Shield Area

A shield is a part of the earth's crust which has not been seriously disturbed since the Precambrian time, though this apparent stability is only relative. The most typical characteristics by which the Precambiran shields are recognized are migmatization and granitisation, indicating a high degree of metamorphism undergone by these rocks.

The shield comprises the oldest geological formations beginning with the Archaeans which has provided a large repository of mineral wealth in India. The other Precambrian formations which contain mineral deposits are Delhis, Cuddapahs, and Vindhyans.

(a) Dharwar System of Karnataka and adjoining states in southern India
(b) Eastern Ghat belt in southern and eastern India
(c) Singhbhum–Gangpur–Bijawar belt east and central Peninsular India

G.S. Roonwal, *Mineral Exploration: Practical Application*,
Springer Geology, DOI 10.1007/978-981-10-5604-8

(d) Sanuar–Sakoli group of central India, and
(e) Aravalli super group north-western Peninsular India

Major Geological Formations of India

Age—estimate at boundary in million of years		
0.010	Recent	Recent Alluvia, Sand dunes, Soils
5	Pleistocene	Older Alluvia, Karewas of Kashmir, and Pleistocene river terraces, etc.
24	Mio-Pliocene	Siwalik, Irrawaddy and Manchhar Systems; Cuddalore, Warkilli and Rajamahendri Sandstones
38	Oligocene-Miocene	Pegu Systems; Nari and Gaj Series
55	Eocene	Ranikot–Laki–Kirthar–Chharat Series; Eocene of Burma
63	Lower Eocene: Upper Cretaceous	Deccan Traps and Inter-trappeans
96	Cretaceous	Cretaceous of Trichinopoly, Assam and Narmada Valley; giumal and Chikkim Series: Umia beds
138	Jurrassic	Kioto Limestone and SpitiShales; Kota–Rajmahal and Jabalpur Series
205–240	Triassic	Lilang System including Kioto Limestone; Mahadeva and panchet series
290	Permian	Kuling System; Damuda System
360	Carboniferous	Lipak and Po Series; Talchir Series
410	Devonian	Muth quartzite
475	Silurian	Silurian of Burma and Himalayas
500	Ordovician	Ordovician of Himalayas
570	Cambrian	Haimanta System/Garbyang Series
1600	Precambrian and	Delhi–Cuddapah and Vindhyan Systems; Dogra and Simla Slates; Martoli Series
2500	Upper Archaean	Dharwar and Aravalli Systems; Salkhala, Jutogh and Daling Series, different types of gneisses
3600	Lower Archaean	

All stages may or may not be present in a given belt. Clear-cut distinctions into the four stages may not be easy in many cases. Cyclic repetition of one or more stages is also recognized.

In the Dharwar of southern India, two stages of the development of a mobile belt are recognized; (i) an initial stage of two cycles and (ii) an intermediate and late stage. Each stage has given rise to typical mineral deposits genetically associated with the intrusive or extrusive rock typical of that stage.

The Himalayan Mobile Belts

Two structural metallogentic episodes are recognized in the Himalayan region. (i) One belongs to the Hercynian mobile belt. The Hercynian mobile belts were intruded by granites and gneisses. During the development of the Himalayan mobile belt, these granites and gneisses were reworked. (ii) In the eugeosynclinal areas of Himalaya–Naga–Lushai belt, ultramafic and mafic intrusive are present. They belong to the lower, middle and upper stages of the development of the mobile belt. In them mineralization, both of magmatic and magmatic metamosphic types, is seen within the intrusives. The associated deposits of chromium, iron and copper can be observed.

The major pre-Himalayan era structural metallogenic belts are as follows:

Dharampur–Bajila–Shishkhani belt	Pb, (Zn, Fe, Cu, Ag, Au, Sb, Ba); Cu, (Pb, Fe, As)
Totan–Dakoti–Basantapur–Aiyur–Pindki belt	These zones occupy the axial portions of the regional folds and are associated with metamorphosed Palaeozoic and Precambrian sediments
Subathu–Ser–Kamag–Kondo belt	
Jari–Manikaran–Uchich–Kotkandi–Jaorinala belt	
Rampur–Benali–Lari–Kammarli, Kara Kunjan	
Narkasi Forest–Prankutiran belt	
Kistwar–Shumahal–Dul belt	
Benihal–Khabrel belt	
Najawan–Kulam–Lashteal belt	

Mineralization has been correlated to the granites which occur nearby.

Such zones occupy the axial portions of the regional folds and are associated with metamorphosed Palaeozoic and Precambiran sediments. Mineralization has been correlated to the granites which occur nearby.

Within the Himalayan mobile belt, the following—metallogenic belts are identified:

Hanle–Chu belt	Show mineralization of Cu, (Fe); Cu, (Co, Ni, Fe, Zn, As)
Dras–Tashgam belt	
Kohima–Ninghthi–Kongal–Nungon belt	
Chakkargaon–Rutland Island belt	

The Platform Areas

The platform areas comprise sedimentary rocks with little of endogenetic mineralisation. The Ironstone-Shale-barren Measure zone of the Gondwanas is not economically significant. However, the clays of Gondwanas, especially the fireclay which occurs along with coal seams, are being mined. Several occurrences of good-grade clays of the Tertiary period are also known.

Cambrian to Recent geological times are represented by various groups of rock formations typically developed in the peninsular and extra-peninsular regions. Such formations have not shown indication of mineralization. The deposits seen in these formations are limestone, salt, gypsum, building stones and occasionally phosphatic nodules.

Several deposits are known to be associated with the lateritic formations and the Quaternaries. In India, the lateritic terrain has broadly been explored. Quaternary formation is yet to be explored. These are of interest to exploration geologists.

Areas of Secondary Mineralization

The areas are zones and belts of primary mineralization where the where the secondary processes have produced economic mineral deposits. The processes of accumulations are weathering and transportation. The two groups of formations of interest are (i) laterite and (ii) quaternary formation.

Lateritie has been rather broadly placed within and above the Tertiary formations in the geological scale. They occur as cappings, occasionally as boulders over rocks which have undergone lateritization. The process of lateritization affects the metal/mineral constituents of host rocks in a way as to concentrate them into economic mineral deposits such as bauxite. Some laterites have yielded valuable mineral deposits such as iron ore, manganese ore, and nickel ore. The important lateritie regions in India and the associated mineral deposits are the following:

Location/Horizon	Metal
Ranchi–Palamau zone	Al, Fe
Bonai–Keonjhar–Singhbhum	Mn, Fe
Bolangir–Patna zone	Mn, Fe
Balaghat–Mandia zone	Al, Fe
Bilaspur–Shahdol zone	Al, Fe
Jabalpur zone	Al, Mn, Fe
North Kanara–Belgaum zone	Al, Mn
Kolhapur zone	Al
Salem zone	Al
Bababudin zone	Al, Fe

(continued)

(continued)

Location/Horizon	Metal
Sambalpur–Kalahandi zone	Al Fe
Chitaldurga–Chickmagalur–Shimoga–Tumkur zone	Mn
Sandur–Bellary zone	Mn, Fe
Goa zone	Fe
Jammu and Poonch zone	Al, Fe
Sukinda zone	Cr, Ni, Co

Lateritization is not always an ore-forming process, the widespread occurrence of important mineral concentrations within lateritic cappings makes it an important rock of interest to the geologists. Once they support mineral deposits like bauxite, clay, manganese and nickel.

Mineralization in Quaternary Formations

Quaternary system is recognized as a time stratigraphic unit, comprise of the youngest sequence of strata including recent sediments being deposited at present. The Quaternary formations of India include glacial deposits, river deposits, desert deposits—sand dunes, formations. It is such formations that diamonds (riverine-accumulation–Panna, tin deposits of Bastar, iron ore, gold, Ranchi, monazite, Kerala are found in the form of floats and fine particles). Quaternary formations have yet to be fully studied.

Mineralizationin Other Formations

Form Cambrian to Pleistocene Age, a large number of formations are recorded in the peninsular and extra-peninsular India. The formations, together with the associated mineral deposits, are listed in the below table.Other geological formations and mineral deposits associated with them

Age of the formation	Name of the formation	Mineral deposits
Cambrian	Salt Range of Himalaya	Rock-salt, gypsum and dolomite, soil
Ordovician	Part of Haimanta system of Himalaya	No mineral deposit noticed
Silurian	–	No mineral deposit noticed
Devonian	Muth Quartzite	No mineral deposit noticed
Carboniferous to Permian	Gondwana	Coal, fireclay, iron ore

(continued)

(continued)

Age of the formation	Name of the formation	Mineral deposits
Triassic	–	Limestone
Jurrassic	–	No mineral deposit noticed
Cretaceous	–	Phosphortic
Upper cretaceous to lower eocene	Deccan trap	Quartz, amethyst, chalcedony, of gem variety

Index

G.S. Roonwal, *Mineral Exploration: Practical Application*,
Springer Geology, DOI 10.1007/978-981-10-5604-8

Printed in the United States
By Bookmasters